CALIFORNIA NATURAL HISTORY GUIDES

INTRODUCTION TO CALIFORNIA MOUNTAIN WILDFLOWERS

California Natural History Guides

Phyllis M. Faber and Bruce M. Pavlik, General Editors

Introduction to

CALIFORNIA MOUNTAIN WILDFLOWERS

REVISED EDITION

Philip A. Munz

Edited by Phyllis M. Faber and Dianne Lake

UNIVERSITY OF CALIFORNIA PRESS

Berkeley   Los Angeles   London

California Natural History Guides No. 68

University of California Press
Berkeley and Los Angeles, California

University of California Press, Ltd.
London, England

© 2003 by the Regents of the University of California

Library of Congress Cataloging-in-Publication Data

Munz, Philip A. (Philip Alexander), 1892–1974
   Introduction to California mountain wildflowers / by Philip A. Munz. —
   New ed.
       p.   cm. — (California natural history guides ; 68)
   Includes bibliographical references (p.   ).
   ISBN 0-520-23635-1 (hardcover : alk. paper) — ISBN 0-520-23637-8 (pbk. :
   alk. paper)
     1. Wild flowers — California — Identification. 2. Mountain
plants — California — Identification. 3. Wild flowers — California —
Pictorial works. 4. Mountain plants — California — Pictorial works. I. Title.
II. Series.

   QK149 .M796 2003
   582.13´09794´09143—dc21

                                                                    2002029142

Manufactured in China
12   11   10   09   08   07   06   05   04   03
10   9   8   7   6   5   4   3   2   1

The paper used in this publication meets the minimum requirements of
ANSI/NISO Z39.48-1992 (R 1997) (*Permanence of Paper*). ♾

The publisher gratefully acknowledges the generous
contributions to this book provided by

the Moore Family Foundation
Richard & Rhoda Goldman Fund
and
the General Endowment Fund of the
University of California Press Associates.

———————

Grateful acknowledgment is also made to
John Game and William T. and Wilma Follette
and
to the California Academy of Sciences

CALIFORNIA
ACADEMY OF
SCIENCES

for their contribution of photographs.

CONTENTS

EDITOR'S PREFACE
TO THE NEW EDITION

California Mountain Wildflowers has introduced thousands to the wildflowers of the mountainous areas of California. Since it was first published in 1963, a number of plant names have been changed, and, in some cases, new information has been obtained. In this revised and updated edition, a number of steps have been taken to make the book current in content and appearance.

The first step was to review the selection of plants included. Most plants that have become rare or endangered since 1963 have been replaced by plants that are more widespread throughout the state. Scientific names for each plant have been updated to conform to the current California authority, the *Jepson Manual: Higher Plants of California,* J. Hickman, editor (University of California Press, 1993). In addition, each plant in this edition has been given a common name using the following sources, listed here in descending order of preference: the *Jepson Manual*; Philip Munz, *California Flora* (University of California Press, 1959); and Leroy Abrams, *Illustrated Flora of the Pacific States* (Stanford University Press, 1923–1960). The rule developed by Munz for hyphenation has been used for all common names: if a plant's common name indicates a different genus or family, a hyphen is inserted to show that the plant does not actually belong to that genus or family. Thus, "skunk-cabbage" is hyphenated because the plant it refers to is not in the cabbage genus nor the cabbage family, but "tiger lily" is not hyphenated because the plant it refers to is in the lily genus, as well as the lily family.

Each plant description has been carefully checked for accuracy and currency. In several cases, taking into account research done in the last 50 years, a description applied to an entire species in the first edition only pertains to a variety or subspecies, or vice versa. Some species have been absorbed into other species, and some have been split into varieties or subspecies. Some varieties or subspecies have even become separate species. All of these changes have been incorporated into the text.

Geographical distribution and elevational information have also been examined and altered whenever new data were available. A new geographical system has been adopted for *Mountain Wildflowers* because of the great number of changes that have resulted from exploration and research in these areas over the years. The distribution data now follows that used in the *Jepson Manual*. Some of the author's original language was out of date: an effort has been made to retain the Munz "flavor" yet to make the new edition readable, entertaining, and informative to today's readers. Some descriptions have been expanded to include more detail to help distinguish a plant from similar ones or to add interest or other helpful hints in identifying a flower.

Dianne Lake has revised the original plant descriptions by Munz and has added new plant descriptions for this edition. The Press is grateful to her for her meticulous care in making these additions and revisions. Many of the lively drawings of Jeanne Janish mentioned in Munz's introduction have been retained. Color illustrations and new design features have been added to make the book more user friendly. Dr. Robert Ornduff wrote an introduction to the plant communities of California's mountain ranges shortly before his untimely death in 2000. He also wrote parallel essays for the other three Munz wildflower books, revised editions of which are also in production. The new *Shore Wildflowers* debuts with this volume, and *Desert Wildflowers* and *Spring Wildflowers* will follow in 2004.

Phyllis M. Faber
August 2002

ACKNOWLEDGMENTS

Most of the drawings in this book were made by Jeanne R. Janish, whose illustrations in many books on western plants are well known and who has an unusual ability to capture the living appearance of a species even when working largely with pressed specimens. A few by Tom Craig and Rodney Cross were made for my 1935 *Manual of Southern California Botany,* which has been out of print for many years.

It is an especial pleasure to express my appreciation of her great help and wise counsel to Susan J. Haverstick given in editing this book as well as the two preceding on California wildflowers and the larger, more technical *A California Flora.*

Philip A. Munz
October 1962

This book is an attempt to present to general readers not trained in taxonomic botany, but interested in nature and their surroundings, some of the wildflowers of the California mountains in such a way that they can be identified without technical knowledge. Naturally, these are mostly summer wildflowers, together with a few of the more striking species that bloom in spring and fall. They are roughly those from the yellow pine belt upward through the red fir and subalpine forests to the peaks above timberline. Obviously, the 286 plants presented cannot begin to cover all that occur in so great an altitudinal range, especially when the geographical limits of the pine belt are considered.

Mention of the pine belt in California mountains will naturally cause you to think of the Sierra Nevada, but of course this belt also extends into the southern Cascade Range (Mount Shasta and Lassen Peak in California) and the Siskiyou Mountains in the northwestern part of the state. As you move southward in the North Coast Ranges through the Yolla Bolly Mountains to Snow Mountain and its environs in Lake County, you may follow the pine belt even into the Santa Lucia Range of the South Coast Ranges. But it extends even farther south, following the Sierra Nevada into the Tehachapi Mountains, Pine Mountain, or Mount Pinos, and the San Gabriel, San Bernardino, and San Jacinto Mountains, all of which have an extensive area, in terms of both altitude and territory, in the pine belt. Even the mountains of San Diego County, such as Palomar Mountain, Cuyamaca Peak, and the Laguna Mountains, have many wildflowers and trees that are common in the Sierra Nevada. Perhaps 100 species in the Sierran pine belt reach San Diego County, another 100 find their southern limit in the San Jacinto and Santa Rosa Mountains of western Riverside County, and 125 more reach the San Bernardino Mountains. In this book I have also included some plants found only in these southern ranges, as well as some confined to the Coast Ranges. I have made no attempt to represent species from the higher desert mountains, such as the White Mountains and

Panamint Range, because for the most part the wildflower seeker does not travel in the desert in summer.

The California Mountains

In general, the mountains of California consist of two great series of ranges: an outer, the Coast Ranges; and an inner, the Sierra Nevada and the southern end of the Cascade Range, including Lassen Peak and Mount Shasta. The Sierra Nevada, an immense granitic block 400 miles long and 50 to 80 miles wide, extends from Plumas County to Kern County. It is notable for its display of cirques, moraines, lakes, and glacial valleys and has its highest point at Mount Whitney at 14,495 feet above sea level. The Cascade Range, on the other hand, is volcanic, with many extinct volcanoes, the highest in California being Mount Shasta at 14,161 feet. The California Coast Ranges are bounded on the north by the Klamath Mountains, of which the Siskiyou Mountains are the best known. The Coast Ranges—several more or less parallel series of outer and inner ranges with intervening valleys—are divided into those north and south of the San Francisco Bay Area, the North Coast Ranges and the South Coast Ranges, respectively. To the south are the Transverse Ranges, where a pine belt extends primarily into the San Gabriel and San Bernardino Mountains, the latter with the highest point in southern California, namely, San Gorgonio Mountain (also called Grayback Mountain) at 11,485 feet. Then, oriented in a more north-south direction, are the Peninsular Ranges (including Santa Ana Peak, San Jacinto Peak, Santa Rosa Mountain, Palomar Mountain, Cuyamaca Peak, and Laguna Peak), with San Jacinto Peak the highest at 10,800 feet.

For the most part, the pine belt receives considerable precipitation, usually more than 25 inches per year, and a large proportion of it is in the form of snow. In the Sierra Nevada, the snowfall may be tremendous—as much as 450 inches per year. Thus, freezing temperatures are common for several months of the year, but summers may be quite warm during the day with dry air and intense sun. Although the California mountains have some summer rain, they do not get nearly as much as do the Rocky Mountains, for instance. It is not surprising then that in many ways the Sierran flora is not as rich as that of some other great mountain ranges of the same latitude and that, although the Sierra Nevada has sufficient elevation and winter for circumpolar plants, the number of species is not nearly so large as in the Rocky Mountains. Much of the montane flora of the California higher mountains has evolved from local sources and local groups of plants adapted to low-moisture conditions instead of being related to plants of more northern localities. There are, of course, some circumpolar or far northern species, such as toad rush *(Juncus bufonius)*, quaking aspen *(Populus tremuloides)*, mountain-sorrel *(Oxyria digyna)*, long-stalked starwort *(Stellaria longipes)*, saxifrage *(Saxifraga mertensiana)*, seep-spring monkeyflower *(Mimulus guttatus)*, and others. But many montane species are quite local, some even confined to a single mountain range or even part of that range, as are high mountain larkspur *(Delphinium polycladon)* and creeping sidalcea *(Sidalcea reptans)* of the Sierra Nevada; bird-footed checkerbloom *(Sidalcea pedata)*, Parish's rock cress *(Arabis parishii)*, and ash-gray Indian paintbrush *(Castilleja cinerea)* of the San Bernardino Mountains; wing-seed draba *(Draba pterosperma)* of the Marble Mountains in Siskiyou County; Trinity Mountains rock cress *(Arabis rigidissima)* of the Trinity Mountains in Trinity and Humboldt Counties; and San Jacinto prickly phlox *(Leptodactylon jaegeri)* of the San Jacinto Mountains in Riverside County.

One of the interesting features of our mountains is that the dry summer makes possible a great differentiation of habitats. We have moist, often grassy streambanks; wet meadows and swampy places; ponds with sandy or muddy shores; dryish flats either exposed or shaded by trees; dry, rocky slopes and points; talus slopes of loose masses of small stones; and even sheer walls of rock with mere crevices for plant growth. All these habitats can and do have different plants even when at the same elevation. Then, too, changes in altitude alone make for very different climates and lengths of growing seasons, with marked effects on the species to be found.

For the most part, this book considers the vegetation of the yellow pine forest and higher areas. This forest begins, mostly, at an elevation of 2,000 to 3,000 feet in the northern parts of the state and runs up to 6,000 or 7,000 feet, whereas in the southern counties it ranges from about 5,000 to 8,000 feet. It is characterized by yellow pine *(Pinus ponderosa)*, sugar pine *(P. lambertiana)*, Douglas-fir *(Pseudotsuga menziesii)*, incense-cedar *(Calocedrus decurrens)*, white fir *(Abies concolor)*, and California black oak *(Quercus kelloggii)* and has a growing season of four to seven months. Above it is found red fir *(Abies grandis)* forest, at 5,500 to 7,500 or 9,000 feet in northern California, and 8,000 to 9,500 feet in the south. Its growing season is three to four-and-a-half months, and its characteristic trees are red fir, Jeffrey pine *(P. jeffreyi)*, western white pine *(P. monticola)*, chinquapin *(Chrysolepis chrysophylla)*, and quaking aspen. Next higher comes the lodgepole forest, from about 8,300 to 9,500 feet and found largely north of the central Sierra Nevada. The growing season is nine to 14 weeks, and the dominant trees are lodgepole pine *(P. contorta* subsp. *murrayana)* and mountain hemlock *(Tsuga mertensiana)*. Above 9,500 feet and poorly represented in southern California, and above 8,000 or 9,000 feet in the more northern parts of the state, is the subalpine forest, our most boreal forest in character. The growing season is only about seven to nine weeks, and killing frost is possible in every month. Characteristic trees are the limber pine *(P. flexilis)*, foxtail pine *(P. balfouriana)*, whitebark pine *(P. albicaulis)*, lodgepole pine, and mountain hemlock. Next come the alpine fell-fields, or the area above timberline, which is above 11,500 feet in the Sierra Nevada and above 9,500 feet in the North Coast

Ranges. Here is found little except perennial herbs, some of them rather woody, scattered or growing among rocks and often forming cushions of low turf.

How to Identify a Wildflower

It is impossible to discuss plants and their flowers without using the names of some of their parts. Some of these terms are defined here. Consult the glossary for other terms that are unfamiliar to you. In the typical flower, we begin at the outside with the sepals, which are usually green, although they may be colored. The sepals together constitute the calyx. Next comes the corolla, which is made up of separate petals or petals grown together to form a tubular, bell-shaped, or wheel-shaped corolla. Usually, the corolla is the conspicuous part of the flower, but it may be reduced or lacking altogether (as in grasses and sedges), and its function of attraction of in-

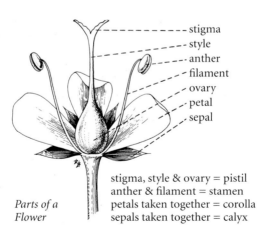

stigma
style
anther
filament
ovary
petal
sepal

stigma, style & ovary = pistil
anther & filament = stamen
petals taken together = corolla
sepals taken together = calyx

Parts of a Flower

A representative flower

sects and other pollinators may be assumed by the calyx. The calyx and corolla together are sometimes called the perianth, particularly where they are more or less alike. Next, as we proceed inward into the flower, we usually find the stamens, each consisting of an elongate filament and a terminal anther where pollen is formed. At the center of the flower is one or more pistils, each with a basal ovary containing the ovules, or immature seeds; a more or less elongate style; and a terminal stigma with a rough, sticky surface for catching pollen. In some species, stamens and pistils are borne in separate flowers or even on separate plants. In the long evolutionary process by which plants have developed into the many diverse types of the present day and have adapted to different pollinating agents, their flowers have undergone very great modifications, and so now we find more variation in the flower than in any other plant part. Hence, plant classification is largely dependent on the flower.

To help you identify a wildflower, either a photograph or a drawing is given for every species discussed in detail, and the flowers are grouped by color. In attempting to arrange plants by flower color, however, it is difficult to place a given species to the satisfaction of everyone. The range of color may vary so completely from deep red to purple, from white to whitish to pinkish, or from blue to lavender that it is impossible to satisfy the writer himself as to whether one color group or another should be used, let alone the readers. I have done my best to recognize the general impression given with regard to color and to categorize the plant accordingly, especially when the flowers are minute and the general color effect may be caused by parts other than the petals. My hope is that by comparing a given wildflower with the illustration it resembles within the color section you think is most correct and then checking the facts given in the text, you may, in most cases, succeed in identifying the plant.

Which Wildflowers Are in This Book

One of the big problems that I faced in writing this book was which species to include. Of course, the title suggests a given limitation: plants of the pine belt and above. In my book *California Spring Wildflowers,* I included many plants from the redwood forests and other coastal areas and from below the pine belt, most of them spring bloomers. *California Desert Wildflowers* includes plants from below the pine belt and on the desert slopes. So, in this book we have the truly montane species, which are primarily summer bloomers. But here I can present only 286 species out of the 1,000 or so that grow in the area. If all were illustrated, the volume would be unwieldy and expensive. I have tried therefore to select plants that are representative of their groups and, in many cases, to mention others in the text. I have presented not only species that are showy and naturally get attention, such as the lilies, but others that are unusual in one way or another and may arouse your curiosity.

I have tried not to be too local in selection of species so the book may be of use in various parts of the state rather than in just the Sierra Nevada, although naturally this area is the most representative of the California mountains and the most widely visited.

Philip A. Munz
Rancho Santa Ana Botanic Garden
Claremont, California
October 1, 1962

INTRODUCTION TO THE PLANT COMMUNITIES OF CALIFORNIA'S MOUNTAIN RANGES

Robert Ornduff

In April 1868, naturalist John Muir stood at the summit of Pacheco Pass southeast of San Francisco Bay and wrote: "Looking eastward...I found before me a landscape...that after all my wanderings still appears as the most beautiful I have ever beheld. At my feet lay the Great Central Valley of California, level and flowery, like a lake of pure sunshine, forty or fifty miles wide, five hundred miles long, one rich furred garden of yellow compositae. And from the eastern boundary of this vast golden flower-bed rose the mighty Sierra, miles in height, and so gloriously colored and so radiant, it seemed not clothed with light, but wholly composed of it, like the wall of some celestial city. Along the top and extending a good way down was a rich pearl-gray belt of snow; below it a belt of blue and dark purple, marking the extension of the forests; and stretching along the base of the range a broad belt of rose-purple; all these colors, from the blue sky to the yellow valley, smoothly blending as they do in a rainbow, making a wall of light ineffably fine." That visual tapestry will never be seen by human eyes again. The level and flowery Central Valley is now mostly given over to agriculture, and the views across it are often obstructed by air pollution; the forests of the Sierra have been logged, replaced by agricultural activities, and altered by the development of numerous communities, outlying ranchettes, and shopping malls. Yet, compared with the agricultural fauna and flora that now occupy the valley floor, the biota and landscapes of California's mountains are still relatively intact.

Although there are few places in California's mountains where human impacts are not evident, there are still many areas where you can see the landscapes and wildflowers that John Muir treasured. The beauty of our mountains results from their diverse geological and historical origins, from the majestic conifers, and from the colorful flowers that grace the forests, meadows, and high alpine regions during summer months. This book introduces you to wildflowers that grow in California's mountains from the yellow pine belt up into the natural rock gardens that occur above the upper limit of tree growth, or timberline. Plants of the desert ranges are not included because, as Philip Munz wrote in his introduction, "the wildflower seeker does not travel in the desert in summer."

California is a land of topographical contrasts: the tip of Mount Whitney reaches 14,495 feet, the highest point in the lower 48 states, and is within sight of the floor of Death Valley, 282 feet below sea level. California's landscape is dominated by mountain ranges that surround the Central Valley and other ranges that separate the southeastern deserts from the Central Valley and the southern coast. All major cities and towns in California are within easy reach of one or more mountain ranges, providing residents and tourists with recreational opportunities throughout the year.

In the north, near the Oregon border, are the Klamath, Siskiyou, and Warner Mountains and the southernmost extent of the Cascade Range, here dominated by Mount Shasta at 14,159 feet. The Coast Ranges begin at the northwestern end of the Central Valley, are interrupted by San Francisco Bay, and continue southward where they join the Tehachapi Mountains and other mountain ranges that rise at the southern end of the Central Valley. The highest peaks in the Coast Ranges reach over 7,000 feet. The southern portion of the state is broken up by numerous ranges—the Transverse and Peninsular Ranges are the largest of these; the Santa Monica, San Gabriel, San Bernardino Mountains and other ranges are somewhat smaller. San Gorgonio Mountain in San Bernardino County reaches 11,500 feet. The deserts have their own set of mountains, with the tallest peak of the White Mountains reaching 14,242 feet.

The dominant mountain range of the state, however, is the Sierra Nevada, Muir's beloved "range of light," which runs northward along the eastern portion of the Central Valley from the Tehachapi Mountains to a point where it meets the southernmost Cascade Range at the northeastern end of the Sacramento Valley. Although Mount Whitney is the highest peak in the Sierra Nevada, at least a dozen other Sierran peaks reach over 13,000 feet; these higher peaks are in the southern portion of this range, but several peaks in the central Sierra reach over 10,000 feet in elevation. The Sierra Nevada occupies about one-fifth of California's land base, but because of its rugged topography, the actual surface area presented by this range is considerably greater. About half the state's native vascular plant species occur in the Sierra Nevada, and of these, over 400 species or other taxonomic entities are found only in this range.

The Sierra Nevada is an enormous tilted block that rises very gradually on the west and drops off sharply to the east of its crest. The drive from Sacramento up to Donner Pass takes hours, but the drive from Donner Pass down to the Nevada border takes only minutes. About 130 million years ago, the Sierran block began to rise, and this uplifting still continues. The crests of Sierran peaks will become even higher with time; occasional earthquakes can cause them to rise a few feet in only seconds. Most of the Sierra Nevada is granitic, but in places, volcanic action has resulted in deposits of ash, pumice, and basalt. Much of the typical Sierran landscape as we know it today was shaped during the Pleistocene glaciations that occurred only a few thousand years ago. During this period, the upper regions were covered by ice sheets that extended down the slopes in river and stream valleys. The rounded appearance of many Sierran peaks and valleys and boulder-strewn granite shields are all results of the action of Pleistocene glaciers. The process of ecological succession is easily seen in the Sierra Nevada: the exposed granite shields such as those in Tuolumne Meadows are gradually being covered by a shallow turf with sedges (Cyperaceae), grasses (Poaceae), and wildflowers. In time, this turf will support a lodgepole pine forest. Grass Lake, near Luther Pass south of Lake Tahoe, was a lake when I first saw it in the 1950s, but in recent years the open water has been covered over by a floating mat of vegetation. In time this lake will be replaced by a meadow, which, in turn, will ultimately be replaced by a coniferous forest.

The Cascade Range extends from southern British Columbia through Oregon and Washington into northern California. The southernmost peak of this range is Lassen Peak (10,457 feet), commonly called Mount Lassen, which last erupted in 1914 to 1917. To the northeast of this southern arm of the Cascade Range is a region called the Modoc Plateau, also volcanic in origin. This high, arid plateau is at the southwestern edge of the enormous Columbia Plateau, which extends northward into Washington and eastward to Utah and Wyoming. In the northeastern portion of the state next to the Nevada border lie the Warner Mountains; these are volcanic in origin, but their volcanic landscape has been obscured to some extent by uplifting due to the continuing action of a fault that runs north to south. As a result, the western slopes of this range are gradual, but the eastern ones are steep, as is true in the Sierra. The highest peak in the Warner Mountains is Eagle Peak, at 9,983 feet.

The Coast Ranges are relatively low, with only a few peaks reaching over 7,000 feet. Because these ranges were not high during the Pleistocene, they were little affected by glaciation. These mountains began to rise at about the same time as the Sierra Nevada. Their complex geological makeup includes rock types such as granite, shale, limestone, basalt, and serpentinite. The Transverse Ranges, unlike most other mountain ranges in North America, run approximately west to east rather than north to south. These mountains are still rising, and although the range is relatively small, it contains one of the most diverse assemblages of rock types in the state. The Peninsular Ranges contain rock types of different origins, some of which are identical to and continuous with those of the southern Sierra Nevada.

The north-south orientation of the Coast Ranges along the immediate coast and of the Cascade Range and Sierra Nevada along the eastern side of the Central Valley influence the climate of the state. Most of the state west of the Sierra Nevada supports a Mediterranean climate, a relatively mild climate with low annual precipitation that falls mostly during winter. The prevailing winds in California are westerly, coming in from the Pacific. They carry moisture-laden air to the interior of the state, but as air rises, it cools, and as it cools, it loses moisture. Thus, the western slopes of the Coast Ranges are wetter than their eastern slopes and the valley floor to the interior, which are in the rain shadow of these ranges. Similarly, the western slopes of the Sierra Nevada are increasingly wetter as you ascend these slopes, though annual precipitation drops at upper-middle elevations and continues to decrease from there up into alpine regions. The eastern slopes of this range are drier and colder than the western slopes, and much of the Great Basin region to the east lies in the Sierran rain shadow.

As you move from north to south in California, the annual precipitation tends to drop off. The mean annual precipitation at Fort Bragg on the coast is 38 inches; at Santa Cruz it is 32 inches; at San Luis Obispo it is 22 inches; and at San Diego it is 11 inches. The mountains to the east of each of these coastal sites receive up to twice their rainfall. The floor of the northern Sacramento Valley receives as much as 30 inches of rainfall per year, but the floor of the southern San Joaquin Valley receives as little as 6 inches per year. The maximum annual precipitation at middle elevations of the western slope of the northern Sierra Nevada is as much as 70 inches per year, whereas corresponding elevations at the southern end of this range may receive only 40 inches per year. Likewise, mean minimum winter temperatures at a given elevation in the Coast Ranges and the Sierra Nevada during winter tend to be lower in the north and higher in the south, and mean maximum summer temperatures may be lower in the north than in the south.

Vegetational patterns reflect these climatic patterns. The western slopes of the Coast Ranges are occupied by plant communities that mostly have higher water requirements than those further inland. As you move from west to east in the northern Coast Ranges, the coniferous forests on their wet western slopes give way to a mixed evergreen forest and eventually to open oak woodlands as you move toward the drier interior. Each of these vegetation types has successively lower water needs. Further eastward were the valley grasslands that once occupied the drier floor of the Sacramento Valley. The coniferous forests of the North Coast Ranges disappear in southern Monterey County, where the precipitation is too low to support them. As you move southward in the Sierra Nevada, the various plant communities that occur on the western slopes gradually move upward in their elevational ranges, a reflection of the changes in precipitation and temperature regimes at a given elevation at different latitudes. Thus, timberline in the southern Sierra Nevada is higher than in the north.

Striking climatic changes occur as you drive up the western slopes of the Sierra. A hypothetical transect in central California gives a mean maximum July temperature on the Central Valley floor of about 76 degrees F, of about 69 degrees F at 4,000 feet, and of 52 degrees F at 10,000 feet. The mean January minima for these same sites are 45 degrees F, 37 degrees F, and 23 degrees F, respectively. The mean annual precipitation at these sites is 20 inches, 70 inches, and 25 inches, respectively, falling exclusively as rain at the lowest site and mostly as snow at the highest site. Thus, the uppermost elevations of the Sierra Nevada are arid, and the little water that is present in the soil is in the form of ice during much of the year.

This book describes and illustrates the common shrubs and herbaceous wildflowers that grow in California's mountains. The lower elevational limit that the author, Philip Munz, has chosen to include is occupied by yellow pine forest. He included plant communities that occur in this forest and above it into alpine regions above timberline. Yellow pine forests form a patchwork with other plant communities wherever they occur and are most extensive on the western slopes of the Sierra Nevada. The yellow pine *(Pinus ponderosa)* forest ranges from about 2,000 to 7,000 feet in elevation in the northern Sierra and from 5,000 to 8,000 feet in the southern Sierra. Its lower elevational limit in the Sierra is marked by about 25 inches of rainfall per year, and its upper limit is marked by 80 to 90 degrees F summer maxima and 22 to 34 degrees F winter minima. Yellow pine forest is often characterized by a mixture of coniferous tree species such as the white fir *(Abies concolor)*, incense cedar *(Calocedrus decurrens)*, sugar pine *(P. lambertiana)*, Douglas-fir *(Pseudotsuga menziesii)*, and giant sequoia *(Sequoia gigantea)* (trees not described in this book). Throughout their wide ranges, these characteristic trees grow with nearly 200 other trees and shrubs and well over 1,000 herbaceous species. Perhaps one-fourth of the native plant species in California grow in yellow pine forest.

Numerous climatic variables in montane plant communities affect the occurrence of plant species. The amount, timing, and form of precipitation (rain versus snow) are all significant. Annual and daily temperature cycles are important; subalpine and alpine regions may experience severe frosts during summer months. The stability, depth, texture, and nutrient status of soils influence plant growth. Winds are important in determining the timberline. The frequency and intensity of fires in forest communities play a role in determining the nature of the plant cover of a region. For example, yellow pine seedlings are intolerant of litter and shade and grow best in cleared areas that receive high levels of sunlight. When frequent light ground fires occur in yellow pine forest, adult trees are undamaged but the undergrowth is cleared, providing conditions that favor the establishment of yellow pine seedlings. If fires are prevented, the shade- and litter-tolerant seedlings of the white fir and incense cedar become established. Eventually these conifers dominate the area, causing heavy shading and litter that prevent reproduction of yellow pine (and several other tree species). Without the conditions created by repeated fires, the rich herbaceous and shrub flora that is typical of the open, parklike yellow pine woodlands will also disappear.

In the Sierra Nevada, yellow pine forests occupy relatively wet, lower and middle elevations that have moderate annual temperatures. Above midelevations, not only do mean annual temperatures continue to drop, but amounts of precipitation also decline. Most of the precipitation in the higher regions falls as snow during winter and in many areas is available for plant growth only during a relatively short period in late spring and early summer. These higher elevations are occupied by subalpine forests, or snow forests. The red fir *(Abies grandis),* white fir *(A. concolor),* whitebark pine *(P. albicaulis),* western white pine *(P. monticola),* lodgepole pine *(P. contorta* subsp. *murrayana),* and mountain hemlock *(Tsuga mertensiana)* are common conifers in this forest. These trees may occur in mixed or pure stands, depending on local environmental conditions. Summer days are generally cool, summer frosts may occur, and soil moisture is mostly the result of snowmelt. Winters are harsh and cold. The upper limit of these forests is marked by timberline, often occupied by picturesquely windblown and stunted individuals of the whitebark pine in the Sierra from Lake Tahoe southward and by mountain hemlock north of Yosemite (the latter species considered by Muir to be "the most singularly beautiful of all California conifers"). Subalpine coniferous forests generally are so dense that few herbaceous plant species can grow in the litter and shade produced by the trees, but mountain meadows, bogs, lake margins, and stream banks support a fine array of colorful wildflowers that are worth finding, such as lilies (*Lilium* spp.), gentians (*Gentiana* spp.), buttercups (*Ranunculus* spp.), and elephant's heads (*Pedicularis* spp.). In subalpine areas with high soil-moisture levels, you can find quaking aspen *(Populus tremuloides)*, a beautiful deciduous hardwood tree that forms enormous groves. In fall, these trees provide a stunning spectacle with their gold and orange leaves.

The eastern slopes of the Sierra Nevada are steeper, drier, and colder and have poorer soils than the western slopes. The plant communities on these slopes often contain mixtures of tree species that often do not grow together on the western slopes. Jeffrey pine *(P. jeffreyi)* is particularly common here. Great Basin plant communities (sagebrush, pinyon pine/juniper, etc.) may occur relatively high on these slopes as well. Plant communities above timberline are alpine communities; they occur from above 6,600 feet in the Klamath Mountains to above 11,000 feet in the southern Sierra Nevada. The upper limit of plant growth in California mountains is about 13,000 feet, so summertime mountain climbers will find wildflowers such as the mountain-sorrel *(Oxyria digyna)* and heuchera (*Heuchera* spp.) and hulsea (*Hulsea* spp.) growing almost to the very tops of our highest peaks. The environmental conditions in these alpine areas are exceedingly harsh. Solar radiation is intense; thus, many alpine plants have reddish or whitish leaves that protect them from damage by ultraviolet radiation. Levels of atmospheric carbon dioxide and oxygen are low. Winds are fierce, so many alpine plants are low or hug the ground. Diurnal temperature variations can be great, with frost possible any night of summer. The growing season is very short: two months or less.

The number of plant species restricted to California's alpine areas is perhaps only 200 species, but about 400 additional species grow in alpine areas, as well as at lower elevations. Over 90 percent of the wildflowers in California's alpine habitats are perennial herbs; generally they are low plants, often forming cushions or rosettes such as the Clemens' mountain-parsley *(Oreonana clementis)*. Annuals are uncommon, probably because of the longer time it takes for annuals to germinate, grow vegetatively, and flower compared to perennials in the same habitat. Because of the low precipitation, alpine plants possess a number of adaptations to aridity, many of which they share with desert plants. Their growth and flowering behavior accommodates the brief growing season. Food in the form of starches and sugars is often stored in fleshy, underground organs. These stored foods allow the plants to grow very rapidly once the snow has melted and daytime temperatures have risen. The flowers of many alpine species are formed during the preceding growing season and are produced shortly after the plant breaks dormancy, sometimes in a matter of days. Optimal temperatures for various metabolic processes are lower than in plants of lower elevations, and the optimal light intensities for photosynthesis may be higher than for plants of lower elevations. Many alpine species are noted for their large and showy flowers; they must compete for a finite population of pollinators during the brief flowering season. These natural rock gardens are spectacular sights when in full flower.

California's alpine wildflowers have different geographical affinities. Many alpine species also occur in the Cascade Range or Rocky Mountains or have close relatives there (e.g., lewisias [*Lewisia* spp.], penstemons [*Penstemon* spp.], and polemoniums [*Polemonium* spp.]). Other alpine wildflowers belong to species or are members of a group of closely related species that occur around the northern hemisphere (e.g., gentians [*Gentiana* and *Swertia* spp.], cinquefoils [*Potentilla* spp.], and saxifrages [*Saxifraga* spp.]). Others are related to high desert species of the adjacent Great Basin, for example, buckwheats (Polygonaceae) and annual monkeyflowers (*Mimulus* spp.). A few alpine species appear to have invaded our alpine regions from lower elevations, where they still grow (e.g., yarrow and phlox).

California's numerous mountain ranges individually offer their own special attractions for wildflower enthusiasts. Many montane wildflower species are known in California only in the northwestern mountains of the state. Most of these reach their southern range limit there but are common in the mountain ranges to the north of the state's boundaries. The highest number of conifer species that grow in one area of North America (17 species) is on Russian Peak in the Salmon Mountains of Siskiyou County. Mount Eddy in the Klamath Mountains supports the largest outcrop of serpentine rock in North America and hosts a number of rare wildflower species. The slopes and meadows of Mount Lassen are occupied by several wildflower species that reach their southern range limit there; these grow side by side with other species that reach their northern range limit there. Butterfly Valley in the northern Sierra Nevada hosts a large number of lilies and orchids and is also a fine place to see large colonies of the peculiar insectivorous California pitcher-plant *(Darlingtonia californica)*. Sierra Valley, just north of Lake Tahoe, offers a diversity of habitats and merits a visit just to see the sea of blue camas that occupies large tracts of the valley floor. The geological mosaic around Winnemucca Lake near Carson Pass supports a rich alpine flora. Hardy hikers who enjoy the rigors of the high country will find the landscapes and wildflowers along the John Muir Trail between Yosemite and Mount Whitney unforgettable. Sequoia and Kings Canyon National Parks are well known for their forests and meadows of wildflowers. The Pebble Plains in Big Bear Valley of the San Bernardino Mountains support what may be the highest concentration of endemic plant species in the entire United States. These are only a few examples of some special places in California where one can see samples of the rich array of mountain wildflowers that grow in our state.

Maps of California

Although not wildflowers in any sense at all, ferns and their allies are plants of woods and fields and interest so many of us that I could not resist putting in a few of the more conspicuous montane forms. **COMMON HORSETAIL (Equisetum arvense)** is one of these. Early in spring, the underground root-

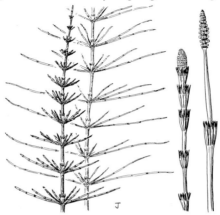

stocks send up simple, pale, flesh-colored, jointed stems to the height of about one foot, with pale sheaths at the nodes bearing eight to 12 lance-shaped, brownish teeth. At the tips of these stems are elongate cones about an inch long formed of whorls of centrally stalked structures, on the underside of which grow many spore-producing organs called sori. The fertile stems are soon followed by green, sterile stems one-half to two feet tall with slender branches in dense whorls, giving a bushy appearance.

LEATHER GRAPE-FERN (Botrychium multifidum) is a stout, rather fleshy fern growing to almost two feet high with a large, triangular, much-divided frond from which rises a fertile spike resembling a cluster of grapes. The spores are contained in roundish cases, called sporangia, on the upper parts of the frond. This fern occurs in moist places and borders of woods below 9,000 feet in the Sierra Nevada, Cascade Range, Modoc Plateau, and central and North Coast Ranges.

BRACKEN FERN, or **BRAKE FERN,** *(Pteridium aquilinum* var. *pubescens)* is a coarse fern with long-creeping, branched underground parts and erect or ascending green fronds that can 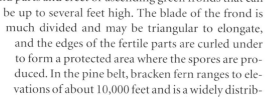 be up to several feet high. The blade of the frond is much divided and may be triangular to elongate, and the edges of the fertile parts are curled under to form a protected area where the spores are produced. In the pine belt, bracken fern ranges to elevations of about 10,000 feet and is a widely distributed groundcover in forests. It can sometimes be mistaken for coastal wood fern *(Dryopteris arguta)* but can be distinguished by the lack of scales on the stem.

LADY FERN *(Athyrium filix-femina* var. *cyclosorum)* is a rather large, robust, tufted plant growing from a short, usually erect, stout rootstock. The fronds are two to five feet high and two to

Bracken fern, or brake fern

Lady fern

three times divided, with numerous minute sori underneath. Occurring below 10,500 feet, this fern is found throughout California except in the Peninsular Ranges, the Central Valley, and desert areas. It, too, can be confused with coastal wood fern (*Dryopteris arguta*), as can be the bracken fern (*Pteridium aquilinum* var. *pubescens*) described above, but the fronds form a diamond shape with the widest branches at the middle and the shortest at the top and bottom. Coastal wood fern is triangular shaped, with its widest branches at the bottom and the shortest at the top.

LACE FERN (*Cheilanthes gracillima*) is a tufted plant growing to about a foot tall. The numerous fronds have dark brown stipes and are twice divided into dull or yellowish green segments that are brownish hairy beneath. This fern grows among rocks from 1,000 to 10,500 feet in most of central to northern California. Somewhat resembling it is coastal lip fern (*C. intertexta*), which has its blades divided three to four times and has bright chestnut brown scales beneath.

Lace fern

Brittle fern, or fragile fern

BRITTLE FERN, or **FRAGILE FERN,** *(Cystopteris fragilis)* is more delicate than the preceding ferns. It has few to several fronds arising from the rootstock, with slender, brittle, smoothish stems and more or

Brewer's cliff-brake

less elongate blades from a few inches to about a foot long. It has thin, pointed frond segments with minute rounded sori beneath. This fern is found mostly in moist, rocky, frequently quite shaded places up to 12,500 feet in most of California.

BREWER'S CLIFF-BRAKE *(Pellaea breweri)* has short-creeping rhizomes covered with twisted brown scales. The fronds are tufted, two to eight inches tall, and divided into thickish, mostly two-lobed, mitten-shaped divisions, or pinnae. The sori beneath are almost covered by the reflexed leaf edge. It grows on exposed, dry, rocky places from 5,000 to 12,000 feet in the Sierra Nevada, Klamath Mountains, Desert Mountains, and Great Basin areas of eastern California. Much like it is Bridge's cliff-brake *(P. bridgesii),* also of the Sierra Nevada, but the pinnae are entire.

American parsley fern

AMERICAN PARSLEY FERN (*Cryptogramma acrostichoides*) is one of the most conspicuous ferns at high elevations. Tufted and almost a foot high, it has two kinds of fronds—one is fertile, or spore bearing, and has narrower and longer pinnules, and the other is sterile and has wider, flat, leathery pinnules. Growing in rocky places from 4,000 to 11,000 feet, it can be found in the San Jacinto and San Bernardino Mountains of southern California, in the Sierra Nevada northward through most of the northern California ranges, and as far north as Alaska and Labrador.

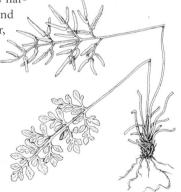

FIVE-FINGER FERN (*Adiantum aleuticum*) has erect fronds one to two-and-a-half feet tall, and the dark, stout stipes are forked above, with each branch bearing several spreading di-

Five-finger fern

Western sword fern

visions four to 16 inches long on the outer sides. The ultimate divisions of the fronds have numerous sori on the undersurface of the truncate tips. This fern grows in moist, shaded places from sea level to 11,000 feet in the South Coast and Transverse Ranges, the northern and central Sierra Nevada, the San Francisco Bay Area, the Cascade Range, and northwestern California to Alaska, Utah, and Quebec.

WESTERN SWORD FERN (*Polystichum munitum*) is a coarse, evergreen fern arising from upright, scaly rhizomes. It has many once-divided fronds two to four feet high, and the pinnae are lance shaped and sharply toothed and have round sori on the undersurface. This fern is common on canyon slopes below 5,000 feet in the southern Sierra Nevada, Cascade Range, and central and North Coast Ranges.

A so-called fern ally is spike-moss, here represented by **ALPINE SELAGINELLA (*Selaginella watsonii*)**. It has branching, prostrate stems one to three inches long with closely overlapping scalelike leaves with minute bristles called setae along the edges. Terminal, erect, green cones contain sori at the bases of their leaves. A locally frequent inhabitant of dry, rocky places between 7,500 and 13,500 feet, this spike-moss occurs in the Peninsular and Transverse Ranges, Sierra Nevada, White and Inyo Mountains, and Trinity Alps of northern California, and to Oregon and Montana.

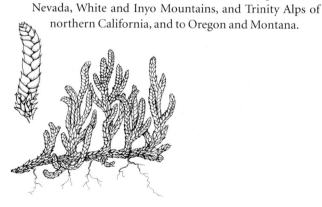

**TIMBERLINE BLUEGRASS (*Poa glauca* subsp. *rupicola)* is included here largely to show what a grass is and to distinguish it from a sedge. The grass family (Poaceae), in general, has hollow stems with solid swellings, or nodes, at the base of each leaf. The leaf ensheaths the stem for a distance and then runs out into an elongate blade. The flowers are minute, without a calyx or corolla, and have only stamens and a pistil enclosed by two leaflike structures, the lemma and palea. These florets occur in small groups, or spikelets, that have two leaflike glumes at their base. The timberline bluegrass is tufted, is less than a foot high, and has flat or folded thin leaves and slender purplish clusters of spikelets. It grows on rocky screes and ridges from 11,000 to 13,000 feet in the central and southern Sierra Nevada and the White and Inyo Mountains.

Several genera make up the sedge family (Cyperaceae), with *Carex* being the largest and containing the true sedges. As compared with grasses (Poaceae), sedges have leaves that are in three rows (not two rows), united leaf sheaths (not split), and a solid (not hollow) stem that is usually triangular. Each flower has a scalelike bract below it. **SIERRA SEDGE (*Carex helleri*)** is a more or less tufted plant that grows to about a foot high and bears terminal clusters of purplish, almost black, bracts below the flowers. It is found on rocky and gravelly slopes from 8,000 to 13,600 feet in the Sierra Nevada, Cascade Range, and Sweetwater Mountains in the eastern Sierra Nevada.

SIERRA ONION (*Allium campanulatum*) is in the lily family (Liliaceae) and belongs to a large group of California bulbous plants that have a strong onion taste and odor when crushed. This species may grow up to a foot high and has two to three flat leaves. At the base of the inflorescence are two oval bracts below the 10 to 50 pale rose flowers loosely arranged in an umbel with the flowers all radiating out from the top of the stem. The individual flowers can be up to one-third inch long. This plant is found on dry slopes in woods from 2,000 to 8,500 feet in the mountains of San Diego County to Oregon, flowering from May to July.

Sierra onion

Swamp onion

SWAMP ONION *(Allium validum)* is a bulbous, green-leaved onion two to four feet high and common in large clumps in wet meadows and on stream banks. The three to six leaves are flat, and the many rose-colored flowers are about one-third inch long. This onion grows from 4,000 to 11,000 feet in the Sierra Nevada and in most of the mountain ranges of northern California from Lake County northward, and also to British Columbia and Idaho. It can be found in flower from July to September.

SCARLET FRITILLARY *(Fritillaria recurva),* also of the lily family, has a stem one to three feet high and usually eight to 10 leaves in one to three whorls around the stem, each with two to five leaves per whorl. The flowers are nodding, scarlet,

Scarlet fritillary

checked with yellow within, and tinged with purple on the outside. This fritillary is found on dry hillsides in brush or woods from 1,000 to 7,000 feet in the Sierra Nevada, Cascade Range, and most northwestern California mountain ranges. It also extends into Oregon and Nevada. Flowering is from March to July.

Another fritillary is **DAVIDSON'S FRITILLARY (*Fritillaria pinetorum*),** with a very glaucous stem four to 16 inches high and four to 20 glaucous leaves that are somewhat whorled around the stem, with the upper leaves shorter than the lower. The flowers are erect or nearly so, purplish, and mottled with greenish yellow. This plant is found on somewhat shaded granitic slopes from 6,000 to 10,500 feet in the Transverse Ranges, Tehachapi Mountains, and Mono County. A species

Davidson's fritillary

resembling it is purple fritillary *(F. atropurpurea)*, which has a more slender, less-inflated stem, more or less equal upper and lower leaves, and no rice-grain bulblets at the base. The nodding flowers are purplish brown, spotted with yellow and white.

SPOTTED CORALROOT *(Corallorhiza maculata),* a member of the orchid family (Orchidaceae), is saprophytic, that is, it lacks chlorophyll and depends on decaying organic matter for its nourishment. The stems are brownish to purplish or even yellow, with whitish sheaths. They become one to two feet tall and have scalelike leaves. The flowers are crimson purple to somewhat greenish, are about one-third inch long, and have a white lip usually spotted with crimson. This plant is found in montane woods below 9,000 feet from San Diego County northward to British Columbia and the Atlantic Coast. It flowers from June to August.

Spotted coralroot

HARTWEG'S WILD-GINGER *(Asarum hartwegii)* belongs to the pipevine family (Aristolochiaceae). It is a stemless perennial herb with creeping rootstocks and heart-shaped leaves two to three inches long. The brownish purple flowers appear near the base of the plant, although they are almost concealed by the leaves. They have no petals, and the hairy sepals are

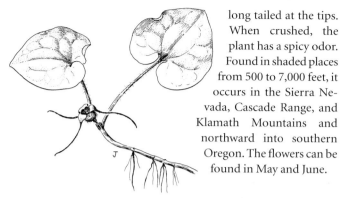

long tailed at the tips. When crushed, the plant has a spicy odor. Found in shaded places from 500 to 7,000 feet, it occurs in the Sierra Nevada, Cascade Range, and Klamath Mountains and northward into southern Oregon. The flowers can be found in May and June.

In the buckwheat family (Polygonaceae), **WATER SMART-WEED** *(Polygonum amphibium* var. *stipulaceum)* is an aquatic perennial with floating lance-shaped or lance-oblong leaves two to four inches long. The small, rose-colored flowers are in terminal, short-cylindrical to ovoid spikes that project up from the surface of the water. It occurs in ponds and lakes below 10,000 feet from San Diego County to Alaska and the Atlantic Coast and blooms from July to September. (See "Whitish Flowers" for a white *Polygonum* species.)

Water smartweed

MOUNTAIN-SORREL (*Oxyria digyna*), also of the buckwheat family, is a low perennial with acidic juice, roundish leaves, and small whorled flowers in compact clusters. The reddish or greenish sepals are scarcely one-eighth inch long. It is remarkable for its wide distribution. Found in rocky places such as crevices in cliffs from 6,000 to 13,000 feet in California, it ranges from the San Jacinto and San Bernardino Mountains northward through the Sierra Nevada and Yolla Bolly Mountains to northern California, and then northward to Arctic America and Eurasia. Its blooming period in California is from July to September.

Another member of the buckwheat family and one of the innumerable kinds of wild buckwheat itself is **OVAL-LEAVED ERIOGONUM,** or **OVAL-LEAVED BUCKWHEAT, (*Eriogonum ovalifolium*).** A dense, almost matted, white-woolly perennial, it has leaves scarcely half-an-inch long and flowering stems only a few inches high. The mostly pinkish or rose flowers are about one-fourth inch long. Several varieties occur, some of them rare, in dry, usually rocky places from 4,000 to 13,000 feet in the San Bernardino Mountains, Sierra Nevada, Cascade Range, Klamath Mountains, and Great Basin areas of eastern California, and also to Oregon and Nevada. They flower from May to July.

PARISH'S ERIOGONUM, or **PARISH'S BUCKWHEAT,** *(Eriogonum parishii)* is a montane species of a large group of annual wild buckwheats from lower and middle altitudes. Erect and growing to about a foot high, it is diffusely branched into a dense mass of very slender ulti-mate branchlets that be-come purplish on aging. It has minute flowers with three outer and three inner pinkish perianth segments. This wild buck-wheat occurs in dry, grav-elly places from 4,000 to 10,000 feet from the southern Sierra Nevada and White Mountains to northern Baja Cali-fornia, flowering from July to September.

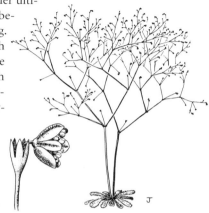

A conspicuous montane representative of the purslane family (Portulacaceae), which is characterized by its two sepals and usually fleshy condition, is lewisia, named for Meri-wether Lewis of the Lewis and Clark expedition to the Pacific Northwest. A species of lewisia, **CLIFF MAIDS (Lewisia cotyledon),** occurs in a number of forms in the Klamath Mountains and Cascade Range to Oregon. The fleshy taproot produces

Cliff maids

several stems up to a foot high, fleshy basal leaves, and many flowers in terminal open clusters. The petals are whitish with red stripes or tinge, often resulting in a rosy appearance, and are half-an-inch long. This lewisia flowers in June and July and occurs from 500 to 7,500 feet. A closely related species that is much rarer and has smaller flowers is Cantelow's lewisia *(L. cantelovii),* found occasionally between 1,000 and 4,000 feet in the same region and also in the northern Sierra Nevada.

BITTER ROOT *(Lewisia rediviva)* is a widespread and variable lewisia. It differs from the above species by bearing large, single flowers, being lower to the ground, and having a very short caudex and stem. It also has six to eight sepals, rather than the two that are usually characteristic of the purslane family. The rose to whitish flowers, with numerous petals from one-half to two inches long, appear in early spring. They are followed by fleshy narrow leaves one to two inches long that form rosettes on the surface of the ground. Bitter root can be found on loose gravelly slopes and rocky places below 10,000 feet from southern California to British Columbia and the Rocky Mountains. It is the state flower of Montana. The flowers bloom from March to June.

Bitter root

YOSEMITE LEWISIA *(Lewisia disepala)* is a rare lewisia similar to bitter root *(L. rediviva),* but it is found only within a very small geographical area. It also has a short, perennial caudex, stems about an inch long, and fleshy, somewhat shorter basal leaves. The flowers are solitary, rose to white, and half-an-inch long or more, but they have only two sepals, unlike bitter root. Yosemite lewisia is found in rocky places from about 6,000 to 11,000 feet, mostly on the sum-

Yosemite lewisia

Western spring beauty

mits bordering Yosemite Valley in the Sierra Nevada. Flowering is in May and June.

WESTERN SPRING BEAUTY *(Claytonia lanceolata),* another member of the purslane family, has one to several stems two to six inches high, with few basal leaves and one pair of stem leaves one to three inches long. The rose to white flowers are one-third to one inch long and vary from three to 15 in number. Western spring beauty is quite variable and is found from 5,000 to 8,500 feet in the San Gabriel Mountains of southern California, the central and northern Sierra Nevada, most of the mountain ranges of northern California, and to British Columbia and the Rocky Mountains. It flowers from May to July.

PUSSYPAWS *(Calyptridium umbellatum),* a different-looking member of the purslane family, is a low plant with a basal rosette of leaves and spreading stems that more or less hug the ground. The dense, terminal clusters of small, white to pink flowers have rosy to white sepals and usually red stamens. The short, thread-like style, wider fruit, and terminal inflorescence distinguish it from other similar species. This plant is found on open rocky slopes, especially talus, from 5,000 to about 14,000 feet in most northern California mountain ranges and below 1,000 feet in a few places in the San Francisco Bay Area. It flowers from May to August.

BLEEDING HEART *(Dicentra formosa),* belonging to the poppy family (Papaveraceae), has a fleshy rootstock with slender stems growing to a foot or more high. The leaves are basal, long stemmed, and very divided. The several nodding flowers are generally rose purple and measure over half-an-inch long. There are four petals: the two outer have spreading tips, the two inner are wing crested on the back. The species is found in damp, somewhat shaded places below 7,000 feet in the Sierra Nevada and Cascade Range, and in the Coast Ranges from Santa Cruz northward to southern Oregon. Two other

Bleeding heart

species, steer's head and few-flowered bleeding heart *(D. uniflora* and *D. pauciflora,* respectively), with only one to three flowers, are also montane and can be found up to 12,000 feet. The flowering period of the three species ranges from March to July. (See "Yellowish Flowers" for another species of *Dicentra.*)

MOUNTAIN JEWELFLOWER *(Streptanthus tortuosus var. orbiculatus)* is in the mustard family (Brassicaceae) and has four sepals, four petals, and a biting or acrid taste. Annual or biennial from leafy rosettes, the stems are erect and freely branched above, with distinctive round to oblong clasping stem leaves and terminal inflorescences of rather purplish flowers. The seedpods are arched and spreading. Several forms occur, but this bushy variety is the most common from

Mountain jewelflower

6,000 to 11,500 feet in the Sierra Nevada and most of northern California. Flowering is primarily from June to September.

Caulanthus, a genus closely related to *Streptanthus,* is represented here by the **CLASPING-LEAVED CAULANTHUS,** or **CLASPING-LEAVED JEWELFLOWER, (Caulanthus amplexicaulis).** The slender stem is

seven to 12 inches high, with broad leaves up to about four inches long. The purple flowers are less than an inch long, and the spreading, curved seedpods are between two and five inches long. This plant is found on dry loose slopes from 3,000 to 9,000 feet in the Transverse Ranges, and a rare variety is found in the San Rafael Mountains. Flowering is from May to July.

PHOENICAULIS (Phoenicaulis cheiranthoides), is a conspicuous plant, especially in fruit, and has no common name. This member of the mustard family has a thick, perennial stem less than a foot long covered with the bases of dead leaves. It produces many pinkish flowers and dense

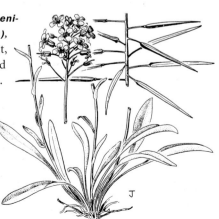

clusters of striking, horizontal, elongate seedpods. This plant is found on dry granitic slopes and benches from 5,000 to 10,700 feet in northwestern California, the Cascade Range, and the Great Basin areas of eastern California. Its northward range is to Washington and Idaho. It can be found blooming from May to July.

CLIFFBUSH (*Jamesia americana* var. *rosea*), a member of the mock-orange family (Philadelphaceae), is a relatively small shrub growing to about three feet high. Its deciduous, opposite leaves are quite hairy and up to about an inch long. The one to 11 rose pink flowers one-fourth inch long are produced in terminal clusters. They are slightly fragrant and have 10 stamens that alternate in length. Found about rocks from 7,800 to 12,000 feet, cliffbush occurs from the central and southern Sierra Nevada eastward through the White, Inyo, and Grapevine Mountains to Nevada. It blooms in July and August.

PURSH'S WOOLLY POD, or **PURSH'S SHEEP POD, (*Astragalus purshii*)** is in a large group known as locoweeds in the pea family (Fabaceae). A perennial with a taproot and tufted or matted stems, it is covered with white wool—even the seedpods are almost con-

Pursh's woolly pod, or Pursh's sheep pod

cealed by the shaggy hairs. The flowers are half-an-inch or longer and range from pink or bright purple to yellowish with purple tips. The species has several forms in California, but as a group, it ranges from about 1,500 to 11,000 feet through much of our montane area from the San Bernardino Mountains northward. Flowering ranges from April to August. (See "Whitish Flowers" for another *Astragalus* species.)

In the geranium family (Geraniaceae), **RICHARDSON'S GERA-NIUM *(Geranium richardsonii)*** is an attractive perennial with one to a few stems growing to about two feet tall or more. The lower leaves are long stemmed and deeply divided into five to seven segments that can be either toothed or cut almost halfway to the midrib; the upper leaves are largely three parted. The flowers are scattered, ranging from white to lavender, with reddish or purplish veins, giving a rose to pink appearance. The five petals are each about an inch long and soft hairy near the base. This perennial occurs in moist places such as meadows from 4,000 to 9,000 feet in the San Jacinto Mountains, Transverse Ranges, Sierra Nevada, and Warner Mountains and to British Columbia. Flowers appear in July and August.

GLAUCOUS SIDALCEA, or **GLAUCOUS CHECKERBLOOM,** *(Sidalcea glaucescens)* belongs to the mallow family (Malvaceae). It has a woody root crown and slender stems one to two feet long. The leaves are deeply lobed or parted into five to seven rather narrow divisions. Flowers are borne along a long flowering stem, the pink to rose petals reaching a length of one-third to two-thirds of an inch. Several similar species occur in the mountains, but this one can be recognized by its very glaucous appearance. It is found in dry, grassy places or open woods from 3,000 to 10,000 feet in the Sierra Nevada and the Cascade Range. It flowers from May to July.

The evening-primrose family (Onagraceae) is usually characterized by four-petaled flowers with a long inferior ovary situated below the petals and sepals. One of the most widely ranging and best-known plants in this family is **FIREWEED** *(Epilobium angustifolium* **subsp.** *circumvagum)*. It is a perennial from underground rootstocks, grows to be one to seven feet tall, and bears long, terminal inflorescences of mostly rose or lilac purple flowers with four petals and long inferior ovaries. The seeds are distributed in the wind by their tuft of hairs, and this plant spreads rapidly into burned and disturbed areas of northern forests, hence, its common name. Found below 11,000 feet, it ranges from San Diego County northward to Alaska, the Atlantic Coast, and Eurasia. Flowering is from July to September.

Fireweed

California-fuchsia, or zauschneria

CALIFORNIA-FUCHSIA, or **ZAUSCHNERIA,** *(Epilobium canum)* is a hairy, often somewhat glandular perennial in the evening-primrose family. Frequently bushy, the stems can reach a length of two feet and are densely leafy, with leaves mostly in pairs, oval to lance ovate, and often half-an-inch or more wide. The brilliant red tubular flowers have the four petals characteristic of the evening-primrose family. Widely spread in dry, rocky places below 10,000 feet, California-fuchsia ranges from San Diego County to southwestern Oregon and flowers in late summer and fall, from August to October.

GLAUCOUS WILLOW HERB *(Epilobium glaberrimum)* is also a member of the evening-primrose family, having its four petals and the ovary beneath the other flower parts. Like the preceding two species, it has seeds bearing tufts of hairs at the tips. There are many species of this genus in mountain meadows and other damp places, this one being a perennial one to two feet high with scaly, wiry rootstocks and a glaucous stem. The lavender to pink petals are up to half-an-inch long and notched. Found at elevations from 2,000 to 12,500 feet, it ranges from the San Jacinto Mountains northward to Washington. It flowers in July and August.

Another *Epilobium* species is **ROCKFRINGE** *(Epilobium obcordatum)*, which has decumbent or matted stems and oval leaves, each branch bearing one to a few flowers near its sum-

Rockfringe

mit. The petals are rose purple and half-an-inch or more long. It is a strikingly beautiful plant growing largely at the base of rocks on dry ridges and flats from 6,000 to 13,000 feet in the Sierra Nevada, Cascade Range, and Modoc Plateau to eastern Oregon and Idaho. It blooms from July to September.

The *Clarkia* genus, of the evening-primrose family, is a very showy group of plants and can be divided into two sections: one with entire, often large petals and the other with narrowed or lobed petals. In the first group, often called godetia, is **LARGE GODETIA** *(Clarkia purpurea* subsp. *viminea).* The flowers range from rose to purple to a deep wine red and often have a red or purple spot near or above the middle. The petals can be over an inch long. The leaves are very narrow, and the flowers are in open, loose clusters. This striking plant can be found below 5,000 feet throughout most of California, blooming from April to July. Two other subspecies occur through much of the state as well—four-spot *(C. purpurea* subsp. *quadrivulnera),* usually smaller flowered and with the stigma never exceeding the

Large godetia

anthers as it does in the other two subspecies, and purple godetia *(C. purpurea* subsp. *purpurea),* which has wider leaves and densely clustered flowers. A similar species found mainly in the Sierra Nevada is Williamson's clarkia *(C. williamsonii).* It can usually be distinguished by its larger, lighter pink petals that can be almost three inches long, and by its usually later blooming period from May to August.

RHOMBOID CLARKIA (*Clarkia rhomboidea*) is in the second section of the *Clarkia* genus. It differs from the godetia group in having a petal that is much narrower below the expanded tip. When in bud, the top of the flowering stem of rhomboid clarkia bends over but becomes erect when the flowers bloom. The rose purple to lavender petals are usually spotted with small, darker purple specks. This plant is found in woodlands and forests in most of montane California below 8,000 feet. It blooms from May to July. Another common *Clarkia* species often encountered that also has the lower parts of its petals narrowed is elegant clarkia (*C. unguiculata*). The upper part of the deep pink petal is almost triangular, and the stem is erect in bud, although the buds themselves are pendent or reflexed. The sepals are fused and all spreading to one side. The plant is often very branched and can reach a height of three feet or more. Rhomboid clarkia is found in wooded areas throughout most of California and blooms from May to June.

The milkweed genus, *Asclepias*, is quite large in California. It is in the milkweed family (Asclepiadaceae), and its members have mostly opposite or whorled leaves, milky sap, and flowers in umbels. **PROSTRATE MILKWEED,** or **SOLANOA, (*Asclepias solanoana*)** has prostrate stems that reach a length of about one foot, with leaves from one to almost two inches long. The flowers are purple outside and white inside, and they have reflexed lobes. This milkweed occurs infrequently on dry serpentine outcroppings from 2,000 to 5,000 feet in the North Coast Ranges from Napa to Trinity Counties, blooming in June.

SIERRA SHOOTING STAR *(Dodecatheon jeffreyi)* is a striking, showy flower in the primrose family (Primulaceae). It is a perennial one to two feet high, and each stem has three to 18 magenta to lavender flowers with a maroon ring at the base and four obtuse stamens. The reflexed petals are one-half to one inch long. The leaves are basal only and three to 20 inches long. It grows in wet places from 2,300 to 10,000 feet in the Sierra Nevada, the Cascade Range, most of northwestern California, and northward to Alaska. A closely related species, mountaineer shooting star *(D. redolens)*, is densely glandular hairy and has five stamens with

Sierra shooting star

pointed tips. It is found from the San Jacinto Mountains to the central Sierra Nevada. Both species bloom from June to August.

SIERRA PRIMROSE *(Primula suffrutescens)* is a true primrose, but unlike our garden primroses, it is woody and has branched, creeping stems. The flower stems are two to four inches high with terminal umbels of small magenta flowers

Sierra primrose

with yellow throats. Sierra primrose frequently occurs under overhanging rocks and on cliffs, mostly from 6,500 to 13,500 feet in the Sierra Nevada and Klamath Mountains. It blooms in July and August.

PIPSISSEWA, or **PRINCE'S PINE,** *(Chimaphila umbellata)* is in the heath family (Ericaceae) and is a low, evergreen, semi-shrubby perennial with thick, shining leaves and rose pink

flowers with five concave, round sepals and petals. It is found on dry, shrubby slopes in woods between 1,000 and 9,500 feet in the San Jacinto and San Bernardino Mountains, Sierra Nevada, and North Coast Ranges. Its distribution also ranges to Alaska and Michigan. It flowers from June to August.

Pipsissewa, or prince's pine

Bog wintergreen, or pink pyrola

BOG WINTERGREEN, or **PINK PYROLA,** *(Pyrola asarifolia)* is also in the heath family. It is an extensively creeping, low perennial herb with leathery broad basal leaves and flowers at the summit of a leafless, somewhat scaly stem. The five pink to rose purple petals are one-fourth to one-third inch long. Found in California in moist shade and woods below 10,000 feet, it occurs in the San Bernardino Mountains, Sierra Nevada, Klamath Mountains, and North Coast Ranges, and northward to Alaska and to the Atlantic Coast. This plant blooms from June to August. (See "Whitish Flowers" for another *Pyrola* species.)

Also in the heath family is a group of plants that lack chlorophyll and get their nourishment from either dead or living organisms. These are called saprophytes and parasites, respectively. **PINEDROPS** *(Pterospora andromedea)* is parasitic on

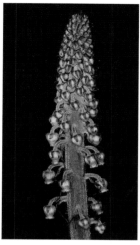

Pinedrops   Snow plant

living root fungi and has purple brown scalelike leaves scattered along a pink to reddish flower stem bearing yellowish, urn-shaped flowers spread out along the stem. It is found in humus in forests below 12,000 feet in the San Jacinto and Tehachapi Mountains, Transverse Ranges, Sierra Nevada, Cascade Range, Klamath Mountains, Warner Mountains, and Coast Ranges and as far as British Columbia and the Atlantic Coast. It flowers from June to August.

SNOW PLANT *(Sarcodes sanguinea)* is a red, fleshy saprophyte in the heath family. It lacks chlorophyll, as does pinedrops *(Pterospora andromedea)* above, but is nourished by decaying, not living, material in the soil. It has many crowded leaves and numerous flowers in a stout, spikelike, terminal inflorescence. The bell-shaped corolla has five broad, red, slightly spreading lobes. Growing in thick humus of forest floors from 3,000 to 10,000 feet, snow plant occurs in the Santa Rosa and San Jacinto Mountains of Riverside County and northward through the Sierra Nevada, as well as in most of the mountain ranges of northern California and into Oregon. The flowers appear between May and July.

AMERICAN-LAUREL *(Kalmia polifolia* subsp. *microphylla)* is also a member of the heath family. It is a low, diffusely branched shrub with opposite leaves having somewhat inrolled edges. The rose purple flowers are in small clusters and almost half-an-inch in diameter. An inhabitant of boggy places and wet meadows from about 7,000 to 12,000 feet, this plant is found in

American-laurel

the Sierra Nevada, Cascade Range, and Klamath Mountains and to Alaska and the Rocky Mountains. Flowering is from June to August.

A final plant in the heath family is **PURPLE MOUNTAIN-HEATHER** *(Phyllodoce breweri),* a low, much-branched, evergreen shrub with narrow, alternate leaves. The rose purple to pinkish flowers are an open bell shape and about one-third inch long, with long stamens extending out of the flower. This plant is found in rocky, sometimes rather moist places from 4,000 to 11,500 feet in the San Bernardino Mountains and Sierra Nevada northward to Magee Peak in the Lassen Peak area. It blooms in July and August.

Purple mountain-heather

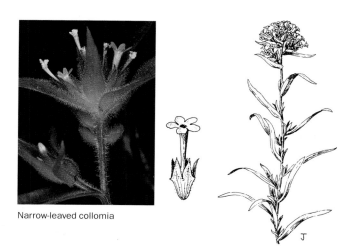

Narrow-leaved collomia

NARROW-LEAVED COLLOMIA (Collomia linearis) is in the phlox family (Polemoniaceae) and belongs to a small group of herbs with alternate leaves and funnelform or salverform flowers. This species is annual and grows to about two feet tall, and its flowers are compacted into heads underlain by several leaflike bracts. The corolla is pink to purplish and about half-an-inch long. The plant is found in dry places, generally between 2,000 and 11,000 feet, from the San Bernardino Mountains to Alaska and Quebec. In California, flowering is from May to August. Another common collomia, but with larger, yellow to orange flowers with blue pollen, is large-flowered collomia (C. grandiflora), which occurs throughout most of California from about 2,000 to 8,000 feet.

The genus *Phlox* is characterized by having the stamens arising from different levels in the corolla tube, as in **SHOWY PHLOX (Phlox speciosa subsp. occidentalis)**. This species has stems that are somewhat woody at the base and can be a foot or more high with pairs of narrow leaves. The few bright pink flowers occur near the ends of the stems and are about half-

Showy phlox

an-inch long. This phlox grows on rocky hillsides and wooded slopes from 1,500 to 8,000 feet in the Sierra Nevada, Cascade Range, Klamath Mountains, and North Coast Ranges and to British Columbia and Montana. It flowers from April to June. (See "Whitish Flowers" for another *Phlox* species.)

Unlike the *Phlox* genus above, the *Gilia* genus, also in the phlox family, has its stamens all arising from the same level.

SPLENDID GILIA (*Gilia splendens*) is an annual, mostly one to three feet high, with a rosette of deeply divided basal leaves. The flowers are in an open terminal inflorescence, pink to pinkish violet, funnel-form, and one-half to one inch in diameter. The species is variable and occurs from 1,000 to 8,000 feet from the San Jacinto and

Scarlet-gilia

San Bernardino Mountains to Monterey County. Flowers appear from May to July.

SCARLET-GILIA *(Ipomopsis aggregata)* is another member of the phlox family. It is the common name of the genus *Ipomopsis* as well as the name of this particular species. It is appropriately named because of its striking, bright red flowers, which usually have some yellow mottling when looked at closely. The tubular flowers, about an inch long, appear in one- to seven-flowered, one-sided clusters on erect stems with pinnately lobed leaves, that is, with the lobes spreading to either side of the midrib. This plant ranges from about 4,000 to 11,000 feet from the central Sierra Nevada northward to British Columbia and to Colorado and Mexico. It can be found blooming from May to September. (See "Whitish Flowers" for another *Ipomopsis* species.)

MUSTANG-CLOVER *(Linanthus montanus),* also a member of the phlox family, is an erect annual up to two feet high. The leaves are remote and deeply divided into five to 11 linear and stiff hairy or bristly lobes. The somewhat funnelform corolla is about an inch long and lilac pink or white with a purple spot on each lobe. This rather lovely species inhabits dry places from 1,000 to 5,500 feet in the central and southern Sierra Nevada. Its flowers appear from May to August. (See "Whitish Flowers" for another *Linanthus* species.)

In the waterleaf family (Hydrophyllaceae), which often has flowers in coiling branches or somewhat rounded clusters, **ROTHROCK'S NAMA,** or **BLUE BALLS,** *(Nama rothrockii)* is a perennial with slender, running, underground rootstocks and coarsely toothed leaves. The purplish lavender flowers are funnelform, about half-an-inch long, and in terminal heads. It is found on dry, sandy flats and benches from 5,500 to 13,000 feet in the San Bernardino Mountains and the central and southern Sierra Nevada. Lobb's nama *(N. lobbii),* shown on the right, is usually found further north and has entire leaves, and flower heads that are both terminal and axillary (rising from the base of the leaf stems). Both species bloom from about June to August.

Related to the preceding *Nama* species is **TURRICULA,** or **POODLE-DOG BUSH,** *(Turricula parryi),* a very glandular,

ill-scented, somewhat woody, erect perennial, branched from the base and with very numerous leaves. It grows to a height of three to eight feet and produces many purplish, funnelform flowers that can be over half-an-inch long. This rather unpleasant plant can cause severe dermatitis in many people. It is occasional in dry places, particularly after fires, and is found below 7,500 feet in the southern Sierra Nevada, South Coast Ranges, Panamint Range, and mountains of southwestern California. It flowers from June to August.

MOUNTAIN-PENNYROYAL *(Monardella odoratissima)* is a branched perennial, somewhat woody at the base, with several stems that can be up to about a foot tall. The leaves can be up to almost two inches long and very aromatic when crushed. The pale purple flowers, slightly more than half-an-inch long, are crowded into heads with colored bracts

below. This perennial is a member of the mint family (Lamiaceae) and therefore has paired, aromatic leaves and two-lipped corollas. It ranges from 1,500 to just over 10,000 feet from the Sierra Nevada and White and Inyo Mountains through most of northern California to Oregon, blooming from June to September.

Mountain-pennyroyal

GIANT HYSSOP, or **GIANT HORSE-MINT,** *(Agastache urticifolia),* also in the mint family, is a tall perennial herb with broad, toothed leaves and several stems branched above. The flowers are rose or violet, about half-an-inch long, and in dense, stemless whorls that may form a more or less continuous spike. This species grows in moist places below 10,000 feet from the San Bernardino Mountains to British Columbia. It flowers from June to August.

RIGID HEDGE-NETTLE, or **RIGID WOOD-MINT,** *(Stachys ajugoides* var. *rigida)* is a widespread member of the mint family. It is a more or less hairy perennial with square stems two to three feet high, and the leaves are usually roundly toothed. The flowers are rose purple or veined with purple and half-an-inch or more long, and the upper lip is much shorter than the lower. This plant has a number of forms and can be found up to 8,000 feet from San Diego County to Washington. Flowering is in July and August.

Giant hyssop, or giant horse-mint

Rigid hedge-nettle, or rigid wood-mint

SCARLET MONKEYFLOWER (Mimulus cardinalis) is a member of the figwort family (Scrophulariaceae), which is similar to the mint family in having two-lipped flowers, but has a different fruit structure. The stems can be round or square, and the leaves can be alternate or opposite. Scarlet monkeyflower is a

Scarlet monkeyflower

Grinnell's penstemon, or
Grinnell's beardtongue

glandular-hairy perennial with erect or decumbent stems from one to almost three feet long, and opposite, longitudinally veined, saw-toothed leaves. The scarlet, sometimes orange corolla is about two inches long. Found on stream banks and moist places below 8,000 feet, it occurs in most of montane California and in adjacent states. It blooms from April to October. (See "Yellowish Flowers" for another *Mimulus* species.)

The large *Penstemon* genus, of the figwort family, is called beardtongue because it has a sterile stamen that is usually strongly bearded. **GRINNELL'S PENSTEMON,** or **GRINNELL'S BEARDTONGUE,** *(Penstemon grinnellii)* is a low, spreading perennial with more or less branched stems that can be from four inches to about three feet long and with pairs of rather broad, usually toothed leaves. The inflorescence is rather open with mostly flesh pink or lavender, sometimes violet, flowers about an inch long. It is to be sought on dry, gravelly, generally granitic slopes from 1,500 to 9,000 feet in the southern Sierra Nevada and the mountains of southwestern California.

BRIDGE'S PENSTEMON *(Penstemon rostriflorus)* is one to three feet tall and woody at the branched base. The inflorescence is quite glandular and composed of tubular, scarlet to vermilion flowers that are an inch or more long and have horseshoe-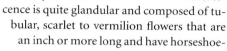shaped anthers. Growing on dry slopes from 5,000 to 9,000 feet, it ranges from the mountains of San Diego County to the central and eastern Sierra Nevada and to Colorado and New Mexico. Often confused with it is another scarlet penstemon, San Gabriel penstemon *(P. labrosus)*, which is not glandular in its upper parts and has divergent anther sacs.

MOUNTAIN PRIDE *(Penstemon newberryi)* is another beardtongue, woody below and matted, with creeping or decumbent stems five to 12 inches long. The rose red corolla is about an inch long, narrow, and only slightly dilated at the throat. Growing in rocky and gravelly places from 2,000 to 11,500 feet, it ranges from northwestern California and the Cascade Range southward through the higher Sierra Nevada to the Tehachapi Mountains. It blooms from June to August. (See "Bluish Flowers" for another *Penstemon* species.)

Red penstemon

Formerly in the *Penstemon* genus, and retaining its former common name, is **RED PENSTEMON** *(Keckiella corymbosa)*. A woody shrub, one to two feet high and densely glandular hairy in the inflorescence, its brick red corolla is an inch or more long and tubular, with the lower lip spreading. It is found on rocky slopes and cliffs below 5,000 feet from Del Norte to Monterey Counties. The flowering season is from June to October.

Great red Indian paintbrush

GREAT RED INDIAN PAINTBRUSH *(Castilleja miniata)* is in another large genus of the figwort family. It has a long tubular corolla with modified lips—the upper lips are elongate and enclose the style and stamens and the lower lips are shorter and somewhat sac-like. The flower is about an inch or more long, and the calyx tips and the leaflike bracts below are scarlet. This showy species grows along streams and in wet places

Dwarf alpine Indian paintbrush

below 11,500 feet in the mountains from San Diego County northward through much of California to British Columbia and the Rocky Mountains. It can be found blooming from May to September.

Smaller and less conspicuous is **DWARF ALPINE INDIAN PAINTBRUSH (Castilleja nana),** which is grayish hairy and less than 10 inches high. The leaflike bract below each flower has three to five lobes and is purplish or yellow green with a green edge. The corolla is about half-an-inch long, and its upper lip is greenish with purple and white edges. This paintbrush occurs in dry, rocky places from 8,000 to 12,000 feet in the Sierra Nevada and blooms in midsummer.

LEMMON'S INDIAN PAINTBRUSH (Castilleja lemmonii) is a purplish red perennial with several simple stems four to eight inches tall and linear, almost undivided, leaves. The bracts have more or less acutely tipped lobes and give the plant its purplish red color. The corolla is three-fourths of an inch long and

glandular hairy on top. The pouch of the lower lip has short white or purplish teeth. This species occurs in moist meadows from 7,000 to 11,500 feet in the Sierra Nevada and Cascade Range. (See "Yellowish Flowers" for another *Castilleja* species.)

Elephant's head

A prominent high elevation plant in the figwort family is **ELEPHANT'S HEAD** *(Pedicularis groenlandica),* which ranges from only three inches to almost three feet tall and is hairless throughout. The leaves on the lower parts are saw toothed, and the red purple flowers are glabrous, with the upper lip of the corolla recurved and narrowed into an upward beak one-fourth to one-half inch long. The outer lobes of the lower corolla lip are ear-like. This plant is occasional in meadows and wet places from 3,000 to 12,000 feet in the Sierra Nevada and northward to boreal America and to the Atlantic Coast. It flowers from June to August.

LITTLE ELEPHANT'S HEAD *(Pedicularis attolens)* is a perennial two to eight inches high and is hairless below but white woolly on the upper stem and in the inflorescence. The leaves are mostly basal and pinnately divided into toothed segments, that is, the segments

Little elephant's head

spread to either side of the midrib. The lavender or pink flowers are in spikelike inflorescences with the beak being less than one-fourth inch long, and the lower lip is fanlike. It is common in meadows and moist places from 5,000 to 13,000 feet in the Sierra Nevada, White and Inyo Mountains, Klamath Mountains, Cascade Range, and Modoc Plateau, and into Oregon. Its blooming period is from June to September.

The last member of the figwort family presented here is **NEVIN'S BIRD'S BEAK** *(Cordylanthus nevinii)*, a slender-stemmed, openly branched annual with yellow roots and linear leaves, the lower leaves having three divisions. The flowers are in heads with one to three purplish, two-lipped flowers about half-an-inch long. This species is found on dry slopes from 5,500 to 8,000 feet in the Piute Mountains of Kern County and from the San Gabriel Mountains of Los Angeles County to the mountains of San Diego County. Similar species occur in northern and central California. The flowering period is from July to September.

In the stonecrop family (Crassulaceae) is **WESTERN ROSEROOT** *(Sedum roseum* subsp. *integrifolium)*, a fleshy perennial from a short, scaly rootstock with several stems one to 12 inches high that bear leaves about half-an-inch long. The flowers usually have four sepals and four petals, the latter dark purple and only about one-eighth inch long. Found in moist, rocky places from 6,000 to 13,000 feet in California, this plant oc-

Western roseroot

curs in the Sierra Nevada, White, Inyo, Klamath, and Warner Mountains, and then to Alaska, Siberia, Colorado, and northeastern North America. It flowers from May to July. (See "Yellowish Flowers" for another *Sedum* species.)

PINK HEUCHERA, or **PINK ALUMROOT,** *(Heuchera rubescens),* of the saxifrage family (Saxifragaceae), is a variable species with a thick caudex, basal roundish leaves, and flowering stems up to a foot long that end in many-flowered, branched clusters. The petals are about one-eighth inch long. It is found in dry, rocky places from 5,000 to 13,000 feet mostly in the Sierra Nevada, but some varieties occur in the Peninsular Ranges and Desert Mountains of southern California. Small-flowered heuchera, or small-flowered alumroot, *(H. micrantha)* is a similar species that occurs in the cen-

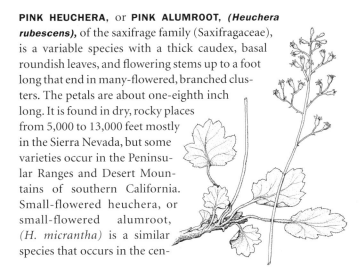

Mountain pink currant

tral and North Coast Ranges as well as in the Sierra Nevada and Cascade Range. It differs from pink heuchera by having a more radial flower and calyx lobes all the same length. Both species can be found blooming from May to August.

A wild currant of unusual beauty because of its rose to rather deep red flowers is **MOUNTAIN PINK CURRANT *(Ribes nevadense)*,** a member of the gooseberry family (Grossulariaceae). It is a slender-stemmed, deciduous shrub three to six feet high and has roundish leaves one to almost three inches wide. The flowers have reddish sepals and shorter white petals, and the glaucous blue black berries are more or less glandular. This wild currant grows in moist places and along streams from Palomar Mountain of San Diego County to southern Oregon, blooming from May to July. In the same genus is Sierra gooseberry *(R. roezlii)*, which also has reddish to purple sepals and white petals but bears one to three spines at each stem node, a characteristic that usually distinguishes the gooseberries from the currants, even though they are in the same genus. The fruit of Sierra gooseberry is red with

Rose-colored meadow-sweet, or rose-colored spiraea

stout prickles and glandular hairs. Mountain pink currant flowers from May to July, and Sierra gooseberry flowers from May to June.

ROSE-COLORED MEADOW-SWEET, or **ROSE-COLORED SPI-RAEA,** *(Spiraea densiflora),* of the rose family (Rosaceae), is a low shrub that has leaves up to about an inch long and a flat-topped inflorescence of pink flowers. The five petals are each only about one-sixteenth inch long, and the flowers have many stamens. This montane spiraea inhabits moist, rocky places from 2,000 to 11,000 feet or may range lower in the northern counties. It is found in the Sierra Nevada northward into the Cascade Range and Klamath Mountains and north-ward to British Columbia, blooming in July and August.

KELLOGGIA *(Kelloggia galioides)* is a slender, perennial herb of the madder family (Rubiaceae) and is related to the whorled-leaved bedstraw. Kelloggia has creeping rootstocks and several stems that can be over a foot high. The paired leaves are gen-erally not more than an inch long. The small, pinkish corolla is funnelform and usually four lobed. Although not a showy plant, kelloggia is common on dry benches and slopes from 3,500 to 9,600 feet from the San Jacinto and San Bernardino Mountains northward to Washington and Idaho. Flowering is from May to August.

CREEPING SNOWBERRY, or **TRIPVINE,** *(Symphoricarpos mollis),* of the honeysuckle family (Caprifoliaceae), is a prostrate or trailing shrub with soft-hairy young twigs and paired round to oval leaves. The bright pink bell-shaped flowers are less than one-fourth inch long and hairy within. The species is found on shaded slopes and ridges below 10,000 feet throughout most of California and into Oregon. It can bloom from April to August, depending upon elevation.

Creeping snowberry, or tripvine

WESTERN TWIN FLOWER (Linnaea borealis var. longiflora), another member of the honeysuckle family, is named for Carl Linnaeus, and portraits of him always show him holding a sprig of it. This plant is a creeping, vinelike perennial with paired leaves and two

Western twin flower

pinkish, funnelform, nodding flowers at the summit of each erect flower stem. Found in dense woods below 8,500 feet, this variety occurs in the northern Sierra Nevada, Cascade Range, Modoc Plateau, and mountains of northwestern California, and to Idaho and Alaska. The species as a whole is circumpolar. It flowers from June to August in California.

In the sunflower family (Asteraceae), what looks like a flower is really a head of many small flowers, or florets. The outer florets (ray flowers) usually simulate petals, and the inner florets (disk flowers) are simply typical, but minute, tubular flowers. **ALPINE DAISY,** or **LOOSE DAISY, (Erigeron vagus)** is a member of a large genus within this family. It is a small perennial with a heavy taproot, crowded basal leaves,

Anderson's thistle

and purple-tinged ray flowers. It is found in scree and rock crevices from 11,000 to 14,100 feet in the central Sierra Nevada and White and Inyo Mountains and to Oregon and Colorado, blooming from July to August. It closely resembles Peirson's aster *(Aster peirsonii)* of similar elevations, which has the leaflike bracts, or phyllaries, below the flower head in more numerous overlapping series.

ANDERSON'S THISTLE *(Cirsium andersonii)* is a biennial or short-lived perennial with a purplish red slender stem one to three feet tall, often having loosely woolly leaves with very spiny teeth and lobes. The showy flower heads have no ray flowers, but the rose purple disk flowers are very slender with long, thin lobes that can resemble ray flowers at first glance. This thistle is found on dry slopes from 2,000 to 10,500 feet in the Sierra Nevada, Klamath Mountains, and Cascade Range. The blooming period is from June to October.

Of the several lessingia (*Lessingia* spp.) that bloom in the mountains of California in late summer, **SIERRA LESSINGIA,** or **SLENDER-STEMMED LESSINGIA, (Lessingia leptoclada)** is probably the one most frequently encountered, at least in the Sierra Nevada. An annual plant, it can range from only a few inches tall up to almost eight feet in height and can be widely branched. It has no ray flowers, but the disk flowers have elongate lobes that often make them look like ray flowers. The heads are usually solitary (though occasionally in clusters) at the tips of branches that are often long and thin. Each head consists of six to 25 pink or lavender to blue disk flowers with unusually long lobes. The phyllaries are multilayered,

Sierra lessingia, or slender-stemmed lessingia

and the reduced upper leaves are glandular and sometimes quite hairy. This lessingia can be found on open slopes from about 1,000 to 6,000 feet through most of the Sierra Nevada and blooms from July to October.

Stephanomeria can be a common sight along mountain roads in late summer with its lovely pink blooms. Several species are found in different parts of California, and it can be difficult to distinguish among them, but the most common is **TALL STEPHANOMERIA _(Stephanomeria virgata)_**. The stiff, branched stem can reach a height of almost 12 feet in extreme cases, but it is usually much shorter. The heads of five to nine pink strap-shaped ray flowers are clustered along the stem branches. Tall stephanomeria occurs on dry open slopes, including disturbed sites, below 7,000 feet throughout most of California and into Oregon and Nevada. It flowers from June to October.

SEASIDE ARROW-GRASS *(Triglochin maritima)*, of the arrow-grass family (Juncaginaceae), is a densely tufted plant one to two feet high. The leaves are fleshy, and the small green flow-ers with six concave perianth segments are in long, ter-minal spikes. The fruit is a cluster of six one-seeded structures, or carpels, that separate when ma-ture. The species occurs in wet, alka-line flats and boggy places, often near hot springs, below 7,500 feet in the San Bernardino Moun-tains, Sierra Nevada, and Coast Ranges and to Alaska, the Atlantic Coast, and Eurasia. The flowers appear between April and August.

Sagittaria latifolia has several common names, including **ARROWHEAD, TULE-POTATO,** and **WAPATO.** A member of the water-plantain family (Alismataceae), it is one to three feet high, with narrow to broad, arrow-shaped leaf blades. The inflorescence is open and branched, and the many white flowers are in three whorls, with staminate (male) flowers above and pistillate (fe-male) flowers below. The petals are half-an-inch or more long. Found at the edges of ponds or slow streams and in meadows below 5,000 feet, it ranges through most of California to British Columbia and the Atlantic states, blooming in July and August.

PARRY'S RUSH, or **WIRE-GRASS,** *(Juncus parryi)* is a member of the rush family (Juncaceae). It is a tufted perennial up to a foot tall, and its leaves are grooved at the base but cylindrical above. The stems are slender and bear one to three mostly brownish flowers, each having six sepal-like parts, three stamens, and a feathery-tipped pistil. This rush grows in rather dry, rocky places from 6,000 to 12,500 feet in the San Bernardino Mountains, Sierra Nevada, and Klamath Mountains and to British Columbia. Other rushes with similar, often greenish flowers and usually a more open inflorescence are common in wet places.

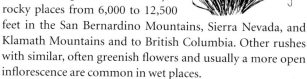

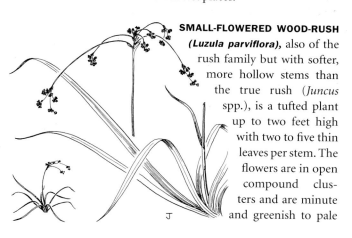

SMALL-FLOWERED WOOD-RUSH (Luzula parviflora), also of the rush family but with softer, more hollow stems than the true rush *(Juncus spp.),* is a tufted plant up to two feet high with two to five thin leaves per stem. The flowers are in open compound clusters and are minute and greenish to pale brown, each with six sepal-like segments. It is found in moist places in woods from 3,500 to 11,000 feet in the Sierra Nevada and most mountain ranges of northwestern California and to Alaska, Labrador, and Eurasia. Other species have flowers in more compact, sometimes spikelike clusters.

WESTERN DWARF MISTLETOE (Arceuthobium campylopodum) scarcely rates as a wildflower, but it is a curious plant that naturally attracts attention with its olive green to yellowish or brownish color. Although in the mistletoe family (Viscaceae), it differs from ordinary mistletoe in having minute, scaly leaves and berries that are on recurved stems instead of being sessile. It has fragile, jointed stems that easily break apart. It is parasitic on Jeffrey pine *(Pinus jeffreyi)*, yellow pine *(P. ponderosa)*, and sometimes Coulter pine *(P. coulteri)* and is found throughout much of California.

In the lily family (Liliaceae), **WASHINGTON LILY (Lilium washingtonianum)** has stems four to nine feet tall and light green leaves in several whorls of six to 12 leaves per whorl. The one to 40 trumpet-shaped flowers are white with small reddish dots. It grows among bushes on dry slopes and flats, generally from 4,000 to 7,000 feet, in the Sierra Nevada, Cascade Range, and Klamath Mountains. It flowers from July to August. A rarer subspecies, purple-flowered shasta lily *(L. washingtonianum* subsp. *purpurascens),* grows at lower elevations in the Klamath Mountains.

A tall, weedy plant in the lily family that is poisonous to livestock is **CALIFORNIA FALSE-HELLEBORE,** or **CALIFORNIA CORN-LILY, (Veratrum californicum).** Forming large clumps

Washington lily

California false-hellebore, or California corn-lily

Plain mariposa

Purple fawn-lily, adder's-tongue, or dogtooth-violet

three to five feet high, it often fills large patches of meadows and stream banks with its clusters of numerous white flowers and its many large, coarsely veined leaves. It grows at elevations up to 11,000 feet and ranges from the mountains of San Diego County northward to Washington and to the Rocky Mountains. Other similar species found in northern California include green false-hellebore, or green corn-lily, *(V.*

viride), with drooping flower branches and green flowers instead of white; fringed false-hellebore *(V. fimbriatum)*, with fringed perianth segments; and Siskiyou false-hellebore *(V. insolitum)*, with woolly ovaries. The flowering period for the four species ranges from June to September.

PLAIN MARIPOSA *(Calochortus invenustus)* seems to be inappropriately named because it is actually a very attractive member of the lily family. It is a bulbiferous plant with slender, erect stems up to about one-and-a-half feet high, with linear leaves that can be up to eight inches long. The white to pale lavender flowers are an inch or more long, often with a purplish spot below the nectar gland near the base of the petal. This species occurs on dry, brushy or grassy slopes and flats from 4,500 to 10,000 feet in the central Sierra Nevada, South Coast Ranges, Tehachapi Mountains, and most of the mountain ranges of southwestern California. It also occurs at slightly lower elevations in the San Francisco Bay Area. It flowers from May to August. (See "Bluish Flowers" for another *Calochortus* species.)

PURPLE FAWN-LILY *(Erythronium purpurascens)* has perhaps a misleading name. A perennial arising from a deep-seated corm with two more or less basal leaves and a flowering stem often about a foot high, the one to six half-inch or longer flowers are actually white with a yellow base when fresh, but they turn purple with age. It also has two other common names that do not do it justice: **ADDER'S-TONGUE** and

DOGTOOTH-VIOLET. It occurs along streams and in meadows or on brushy or forested slopes from 5,000 to 9,000 feet in the northern and central Sierra Nevada and the Cascade Range. It blooms from May to August. Glacier-lily *(E. grandiflorum),* a golden yellow species, can be found in the North Coast Ranges and Klamath Mountains, blooming from June to July.

FALSE SOLOMON'S SEAL *(Smilacina racemosa),* also of the lily family, has a stout rootstock, an erect stem one to three feet high, and broad, mostly clasping leaves. The many small, white flowers are in a branched inflorescence, whereas in the related slim Solomon's seal *(S. stellata),* the fewer flowers are in a simple, more open cluster on a shorter, more slender stalk. Both species are found in shaded woods, often in moist places, mostly below 8,000 feet through much of California to British Columbia and the Atlantic Coast. Both bloom from March to May.

False Solomon's seal

White-flowered bog-orchid

**WHITE-FLOWERED BOG-ORCHID
(Platanthera leucostachys)** belongs to the orchid family (Orchidaceae) and has fleshy, tuberlike roots. The stem can be from one-half to over three feet tall, with scattered green leaves and a terminal spike of white flowers, each about half-an-inch long. This attractive, fragrant orchid is frequent in wet and springy places below 11,000 feet in the mountains from San Diego County northward to British Columbia. Flowering is from May to August.

WESTERN BISTORT (Polygonum bistortoides), of the buckwheat family (Polygonaceae), is a perennial that has several erect, slender, simple stems one to two feet high and basal leaves. The white flowers are in terminal, thick, cylindrical spikes one-half to two inches long, and each flower has six perianth segments about one-fifth inch long. The plant is common in wet meadows and along streams, mostly between 5,000 and 10,000 feet, in the San Jacinto and San Bernardino Mountains, Sierra Nevada, and North Coast Ranges and to Alaska and the Atlantic Coast. It is also found at very low elevations in a few places along the northern coast as far south as Marin County. Its blooming period is June to August. (See "Reddish Flowers" for another *Polygonum* species.)

Western bistort

Wild buckwheat (*Eriogonum* spp.) is an immense genus in California and is here represented by **LOBB'S ERIOGONUM** *(Eriogonum lobbii)*. With a stout, woody, few-branched caudex, it has tufted rosettes of leaves with densely woolly undersides. The flowering stems are often more or less flat on the ground and bear white to rose flowers. It is found on 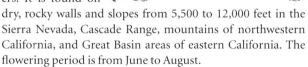 dry, rocky walls and slopes from 5,500 to 12,000 feet in the Sierra Nevada, Cascade Range, mountains of northwestern California, and Great Basin areas of eastern California. The flowering period is from June to August.

ROCK ERIOGONUM *(Eriogonum saxatile)* is a somewhat different type of wild buckwheat and illustrates to some extent the tremendous array of forms this remarkable group assumes

Lobb's eriogonum

in California. Also a perennial, it has densely white-felted leaves; slender, ascending flower stems; and flowers one-fourth inch long that can be white, rose, or yellowish. It can be sought on dry, rocky slopes and ridges, generally from 3,000 to 11,000 feet, in the southern Sierra Nevada, Desert Mountains, and mountains of central and southwestern California. It flowers from May to July. (See "Reddish Flowers" for more *Eriogonum* species.)

In the goosefoot family (Chenopodiaceae), the flowers are minute, greenish or membranous, and without petals. Plants such as the garden beet and spinach, as well as desert-holly, belong to this family. **FREMONT'S GOOSEFOOT**, or **FREMONT'S**

PIGWEED, *(Chenopodium fremontii)* is an erect annual up to two feet high, with triangular leaves greenish above and whitish beneath. It is met along trails and similar disturbed places between 2,000 and 10,000 feet from the mountains of southern California to the Sierra Nevada, in the Great Basin areas of eastern California, and to British Columbia. The flowers appear between June and October.

The purslane family (Portulacaceae) is fleshy, and its flowers usually have only two sepals. **TOAD-LILY *(Montia chamissoi)*** is a perennial with slender, more or less buried runners that produce bulblets. The stem is from one inch to a foot tall and has several pairs of leaves and a cluster of two to eight flowers with white to pinkish petals about one-fourth inch long. It is found in wet places in meadows and along streams from 4,000 to 11,000 feet in the Transverse and Peninsular Ranges, Sierra Nevada, Cascade Range, Klamath Mountains, North Coast Ranges, and Great Basin areas of eastern California and to Alaska and Minnesota. It blooms from June to August.

Toad-lily

LONG-STALKED STARWORT *(Stellaria longipes)*, with its paired leaves and five-petaled flowers, belongs to the pink family (Caryophyllaceae). It is a somewhat tufted perennial growing from creeping rootstocks. The stems grow to about 14 inches tall and have long, narrow leaves. The one to seven flowers have cleft petals that are about one-fourth inch long. It is a dainty little plant found in moist places from 4,000 to 10,500 feet in the San Bernardino Mountains, Sierra Nevada, Cascade Range, Klamath Mountains, North Coast Ranges, and Great Basin areas of eastern California. It also ranges to Alaska and the Atlantic Coast. The flowering period is May to August.

Another member of the pink family is **LEMMON'S CAMPION,** or **LEMMON'S CATCHFLY,** *(Silene lemmonii),* a slender-stemmed perennial that grows to over a foot tall and is glandular in the upper parts although not sticky as some other species of this genus. The mostly nodding flowers have petals

Lemmon's campion, or Lemmon's catchfly

that are divided into four linear lobes and sepals fused together into a tube. It is common in open woods from 3,000 to 9,000 feet from the Cuyamaca Mountains of San Diego County northward to Oregon. It can be found blooming from June to August.

SARGENT'S CAMPION *(Silene sargentii)* is a tufted, glandular perennial about four to six inches high. It has a somewhat tubular calyx and a white corolla with two-lobed petals, each lobe having a small lateral tooth. It is a plant of rock crevices and similar places at altitudes from 8,000 to 12,000 feet and is found in the Sierra Nevada and the Sweetwater and White Mountains. Flowers appear in July and August.

In the buttercup family (Ranunculaceae) is **BANEBERRY** *(Actaea rubra),* a perennial herb one to almost three feet tall, with stem leaves made up of broad, incised, and toothed leaflets. The flower clusters are dense when in flower, but elongate in fruit, the small white petals soon falling away. The fruits are red or white, rather persistent, and about one-fourth inch long. Baneberry grows in rich, moist woods below 10,000 feet in the San Bernardino Mountains, Sierra Nevada, and Cascade Range and in the Coast Ranges from San Luis Obispo County northward. Flowering is in May and June.

Baneberry

A number of varieties of **WATER BUTTERCUP (*Ranunculus aquatilus*)** occur in California. An aquatic perennial with submerged or floating stems, it has divided leaves that are generally submerged and occasionally a few lobed, floating ones as well. The five white petals are about one-fourth to one-half inch long. The plant is found in ditches, slow streams, and ponds below 10,000 feet and ranges throughout California to Alaska, the Atlantic Coast, and Europe. Flowering is from April to July depending on elevation. (See "Yellowish Flowers" for more *Ranunculus* species.)

SLENDER-SEPALED MARSH-MARIGOLD (*Caltha leptosepala var. biflora*) is also in the buttercup family. It has three to eight separate pistils and is a fleshy perennial with large, simple, mostly basal leaves and solitary flowers with white sepals and no petals. It is found in marshy and boggy places from 3,000 to 10,500 feet in most of the Sierra Nevada, Cascade Range, Modoc Plateau, and North Coast Ranges, as well as in southern Oregon. Flowers appear from May to July.

Slender-sepaled marsh-marigold

COVILLE'S COLUMBINE *(Aquilegia pubescens),* another member of the buttercup family, has a stem up to two feet tall with the basal and lower leaves once to twice ternately lobed (divided into three segments) and the upper leaves more deeply three lobed. The erect flowers are white to cream, and each petal has a long, hollow, nectar-bearing spur that attracts the sphinx moth, the plant's pollinator. It is found at higher elevations from about 9,000 to 12,000 feet, particularly on talus and in rocky places in the Sierra Nevada. It blooms from June to August.

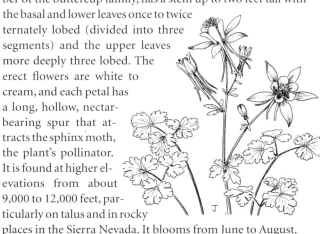

DRUMMOND'S ANEMONE *(Anemone drummondii)* is another perennial in the buttercup family. One to several soft-hairy stems grow up to about a foot tall from a stout root crown.

Drummond's anemone

The leaves are also quite hairy and divided into linear segments. The white or bluish-tinged sepals are about half-an-inch long; petals are lacking. Occurring on talus and gravel or rocks from 4,000 to 10,600 feet, this anemone occurs in the Sierra Nevada, Cascade Range, and Klamath Mountains, and also northward to Alaska and Alberta. It flowers from May to August.

FENDLER'S MEADOW-RUE *(Thalictrum fendleri)* is also a member of the buttercup family. It reaches two to six feet in height and has leaves divided two to four times into broad leaflets. Staminate and pistillate flowers are on separate plants and are greenish, with four to seven deciduous sepals and no petals. The pistils are one seeded. There are two varieties: in *T. fendleri* var. *fendleri,* the upper leaves are glandular and somewhat hairy on the underside and the fruit has two to three ribs on each  side; in *T. fendleri* var. *polycarpum,* the upper leaves are glabrous on the underside and the fruit has only one rib per side. Found mostly in damp places below 10,000 feet, both varieties range from the mountains of San Diego County northward to Oregon and Wyoming and are also found in Texas. They bloom from May to August.

The mustard family (Brassicaceae) can be recognized by its four petals and six stamens, two lower than the other four. **BREWER'S BITTER-CRESS *(Cardamine breweri)*** is one of the plants found in this family. A perennial with creeping rootstocks, it has a stem that reaches a height of about two feet. The leaves can be either entire or divided into three to five oval lobes, and the flowers have white petals about one-fourth inch long, followed by linear seedpods that can be up to an inch long. This plant grows along streams below 11,000 feet in the San Bernardino Mountains, western Transverse Ranges, Sierra Nevada, Cascade Range, and northernmost mountains of northwestern California, thence to British Columbia and Wyoming. Flowering is from May to July.

HOLBOELL'S ROCK-CRESS (*Arabis holboellii* var. *retrofracta*) is another plant of the mustard family and a member of the large genus *Arabis,* which has many species in America and

Holboell's rock-cress

Eurasia. This rock-cress has stems from eight inches to almost three feet tall and whitish to purplish pink flowers that produce long, slender, almost straight, reflexed seedpods. In many related species, the pods are strongly arched and often spreading. It is found in dry, rocky places from 1,800 to 8,000 feet in the mountain ranges of northwestern California, Cascade Range, Sierra Nevada, parts of the San Francisco Bay Area, and Great Basin areas of eastern California. It flowers from May to July.

Another quite different rock-cress is the little **BROAD-SEEDED ROCK-CRESS (Arabis platysperma),** which is less than a foot tall and has entire leaves. The flowers have spoon-shaped petals, and the seedpods are flat, ascending, and from one to almost three inches long. This rock-cress grows on dry, rocky flats and slopes from 4,000 to 12,000 feet in the Sierra Nevada, North Coast Ranges, Cascade Range, and Great Basin areas of eastern California. It flowers in June and July.

ROUND-LEAVED SUNDEW (Drosera rotundifolia) is an insectivorous plant in the sundew family (Droseraceae), which also includes the famous Venus flytrap of North Carolina. The leaves are in a spreading rosette, with their upper surfaces clothed with tentacle-like, gland-tipped, reddish hairs that bend over and entrap insects. The small flowers are at the top of a two- to 10-inch-tall stem and have five white or pink petals that are soon shed. It is a plant of cold, wet bogs and swamps around the world. In California it is found below 6,500 feet in the Sierra Nevada, Cascade Range, and mountains of northwestern California. It blooms from July to August.

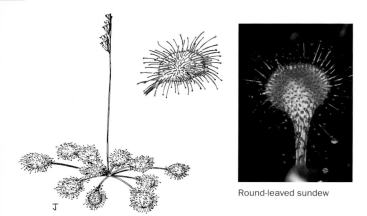

Round-leaved sundew

In the saxifrage family (Saxifragaceae), the floral parts are somewhat united into a tube that may be partly connected to the ovary. **BROOK SAXIFRAGE** *(Saxifraga odontoloma)* is a perennial with roundish, coarsely toothed leaves and flower stems a foot or more in height. The flowers have reflexed sepals and round to elliptic white petals with two yellow dots at the narrowed base. It is found on moist stream banks below 5,000 feet in the San Bernardino Mountains, Sierra Nevada, Cascade Range, Klamath Mountains, and North Coast Ranges and to Washington and the Rocky Mountains. Flowering is from July to August.

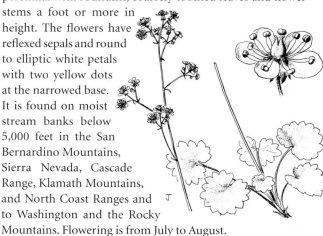

PYGMY SAXIFRAGE *(Saxifraga rivularis)* exemplifies quite a different type of saxifrage. It is also a perennial, but it forms small tufts only two to four inches high. The lower leaves are reniform (kidney shaped) in outline, mostly with three to five lobes, and the upper leaves are much reduced. The white petals are oblong spatulate. It is found in damp, shaded places around overhanging rocks from 11,000 to over 14,000 feet in the Sierra Nevada and also occurs in the Rocky Mountains. The flowering period is from July to August.

Still another member of the saxifrage family is the odd little **BREWER'S MITREWORT _(Mitella breweri)_,** which has small, greenish petals divided into linear lobes. The leaves are basal and roundish. This species occurs on damp, shaded slopes from 5,000 to 10,500 feet in the Sierra Nevada, Cascade Range, and Klamath Mountains and also reaches British Columbia and Montana. Other related species with whitish

Brewer's mitrewort

Also in the rose family, and related to the cinquefoils (*Potentilla* spp.), is **DUSKY HORKELIA (Horkelia fusca subsp. parviflora)**, a slender-stemmed perennial with dark green, glandular-hairy leaves divided into five to 10 pairs of leaflets. The

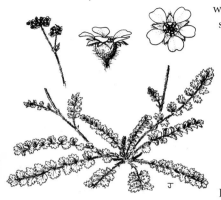

white, spatulate petals are scarcely one-sixth inch long, and the multiple stamens are widened at the base. This horkelia grows in open places from 3,000 to 11,000 feet in the Sierra Nevada, northern Cascade Range, eastern Klamath Mountains, and parts of the Modoc Plateau, then to Washington and Wyoming. Closely related forms and species are common in much of California and are frequently quite aromatic. This plant flowers in July and August.

MOUSETAIL IVESIA (Ivesia santolinoides) is also in the rose family. It is an attractive, perennial herb with a basal rosette of

thin, elongate, silvery-silky leaves. The slender, suberect stems are up to a foot tall, and the many small white flowers give a baby's-breath effect. It is found on dry, gravelly slopes and ridges from 9,000 to 12,000 feet in the San Jacinto Mountains, Transverse Ranges, and Sierra Nevada, blooming from June to August.

A low, shrubby plant in the rose family is **MOUNTAIN MISERY (Chamaebatia foliolosa),** with many leafy branches and glandular, heavy-scented young twigs that soon exfoliate. The leaves are viscid and much divided. The flowers are densely hairy inside near the base and have round, white petals one-fourth to one-third inch long. It is common as a ground cover in open forests between 2,000 and 7,000 feet and ranges from the Cascade Range to the Sierra Nevada. The flowers appear between May and July.

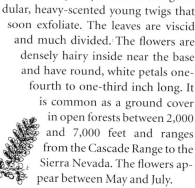

A rather tall and erect shrub in the rose family is **BITTER CHERRY (Prunus emarginata).** It is deciduous in winter and grows to a height of five to 20 feet. The flat-topped clusters of three to 12 small, white flowers produce bright red bitter fruits. It can be found on dryish rocky ridges or dampish slopes below 9,000 feet and ranges in the mountains from southern California to British Columbia and Idaho. It flowers in April and May. Sierra plum *(P. subcordata)* and western choke-cherry *(P. virginiana* var. *demissa)* are also frequently encountered in the mountains of California. The former has more open clusters of one to seven and usually larger flowers that produce yellow to dark red fruits about an inch long, which are edible to humans. It flowers from March to May. The latter has elongate clusters of many small, white flowers that produce many small, dark red to black fruits that contain cyanides and should not be eaten. It flowers from May to June.

Bitter cherry

THIMBLEBERRY *(Rubus parviflorus)* is one of the several mountain berry bushes of the rose family that produce berries edible to humans. It is deciduous, mostly three to six feet high, without prickles, and with shredding, reddish bark in age. The leaves are four to six inches wide and palmately lobed into five segments, spreading like the fingers from the palm of a hand. The white to pink flowers are one to two inches across, and the red hemispheric berry is edible, but bland. This plant is found in open woods and canyons below 8,000 feet from the mountains of San Diego County to Alaska, and its flowering period ranges from March to August. Two other berry bushes with delicious edible fruit are California blackberry *(R. ursinus)* and blackcap raspberry *(R. leucodermis)*. In both

Thimbleberry

plants, the leaves are generally divided into three leaflets (although they are occasionally just deeply lobed, not divided, in California blackberry). The leaves are green on the underside in California blackberry, and the berry is a deep black. In blackcap raspberry, the leaves are white on the underside, and the fruit is red purple to almost black. Both plants have prickles, unlike thimbleberry. They flower between March and July and generally have mature fruit about a month after they flower.

MOUNTAIN-ASH *(Sorbus scopulina)* is a deciduous shrub in the rose family with thick, reddish bark and leaves of 11 to 13 leaflets. The flat-topped inflorescence has 80 to 200 flowers, each slightly less than half-an-inch in diameter. The orange to scarlet fruits (much like tiny apples) are about one-third inch in diameter. The species is occasional in canyons and on wooded slopes between 4,000 and 9,000 feet in the Sierra Nevada, Cascade Range, Klamath Mountains, North Coast Ranges, and Modoc Plateau, thence to British Columbia and the Rocky Mountains. It flowers in June.

The pea family (Fabaceae) is one of the largest and most important plant families used as a source of food, ornamentals, and lumber. It is noteworthy for its very large genera such as *Trifolium*, or clover. A common perennial clover in the mountains is **LONG-STALKED CLOVER** *(Trifolium longipes)*. It has several varieties, but the flower heads are usually sessile or very short stemmed at the tops of long terminal or subterminal flower stems that are often bent or curved at the tip. The small,

white flowers are often tinged
with purple, and the calyx
has lanceolate to bristle-
like lobes. The leaves
consist of three long,
thin leaflets. It is found
in moist places below
10,000 feet from the
San Jacinto Moun-
tains through the
Sierra Nevada to the
mountains of northern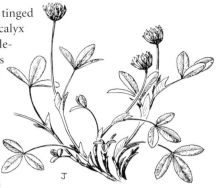
California, and then to Washington and Idaho. It also occurs
in the Great Basin areas of eastern California. The blooming
period is from June to September.

CARPET CLOVER *(Trifolium monanthum)* is another montane
clover and also perennial, but it has very slender stems and
one- to three-flowered heads of narrow, whitish florets. It also
has a number of varieties but in general is found in grassy,
moist places from 5,000 to 12,500 feet in the San Jacinto, San

Carpet clover

Bernardino, and San Gabriel Mountains, South Coast Ranges, Sierra Nevada, and Cascade Range. It blooms from June to August. A much larger species is cow clover *(T. wormskioldii),* which has coarser growth and large heads of white to pink flowers half-an-inch long. Cow clover can be found in bloom from May to October.

One of the largest genera in the pea family in North America is made up of the locoweeds and rattleweeds (*Astragalus* spp.). Some species of *Astragalus* are called locoweed because they poison livestock, and others are called rattleweed because they have dry, inflated pods in which the seeds rattle about in the wind. **BOLANDER'S LOCOWEED** *(Astragalus bolanderi)* is a perennial that grows up to one-and-a-half feet in height. Its leaves are divided into 13 to 27 leaflets. The several flowers are in loose clusters, and each flower is about one-half to two-thirds of an inch long. It is found in dry, rocky and sandy flats and meadows from 5,200 to 10,000 feet in the

Bolander's locoweed

northern and central Sierra Nevada and blooms from June to August. (See "Reddish Flowers" for a pink *Astragalus* species.)

Perhaps California's showiest shrubs are the California-lilacs (*Ceanothus* spp.), of which there are over 40 species in the state. These shrubs belong to the buckthorn family (Rhamnaceae) and have small flowers with flattish, central disks. Many are beautiful, but no such claim can be made for

Tobacco brush

TOBACCO BRUSH *(Ceanothus velutinus)*. Yet it is a conspicuous, spreading, round-topped, evergreen shrub with dark green leaves varnished above and paler beneath. The flower clusters are one to two inches long. This plant occurs on open wooded slopes below 10,000 feet and ranges from the Sierra Nevada through most of northern California and to British Columbia and South Dakota. It blooms from April to July.

Another conspicuous and common species of *Ceanothus* is **MOUNTAIN WHITETHORN,** or **SNOW BUSH *(Ceanothus cordulatus)*.** It is a low, spiny, grayish, glaucous shrub and intricately branched. The alternate leaves are oval to elliptical and three veined from the base. The flowers are in dense, white

Mountain whitethorn, or snow bush

clusters less than two inches long. Found on dry, open flats and slopes from 3,000 to 9,500 feet, it occurs in the San Jacinto, San Bernardino, and San Gabriel Mountains, Sierra Nevada, North Coast Ranges, and Klamath Mountains and northward to Oregon. It blooms from May to July.

Also in the *Ceanothus* genus is **DEER BRUSH (Ceanothus integerrimus)**. It has several named varieties but is generally a loosely branched shrub, three to 12 feet tall. The flowers are usually white but can sometimes be bluish or pink. It is found on dry slopes and ridges between 1,000 and 7,000 feet from southern California to Washington and blooms from May to June. A different *Ceanothus*, which has opposite, thicker leaves with sunken white pits on the underside rather than the smoother green surface of the preceding species, is buck brush *(C. cuneatus),* which is in the same group as squaw carpet *(C. prostratus)* but is taller and erect and has white flowers. It can be found in dry, mostly chaparral areas throughout most of California, blooming from March to May. (See "Bluish Flowers" for another *Ceanothus* species.)

In the evening-primrose family (Onagraceae), a small-flowered but common and often noticeable annual is the slender-stemmed **DIFFUSE GAYOPHYTUM** *(Gayophytum diffusum).* It has alternate, entire, narrow leaves and scattered flowers with petals up to about one-sixth inch long. The ovary is beneath the other flower parts (inferior), and the four petals are white to pinkish in age. Found in dry, open places from 2,500 to 12,000 feet, it can be expected in the mountains from San Diego County northward to Washington. The flowering period is June to August.

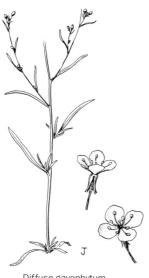

Diffuse gayophytum

Another member of the evening-primrose family but one that has a reduced flower (two sepals instead of four, two petals, two stamens) is **ENCHANTER'S-NIGHTSHADE** *(Circaea alpina subsp. pacifica)*. It is a rather low perennial with a simple erect stem, a few pairs of thin leaves, and minute flowers in long, terminal clusters. The fruit is nutlike and covered with hooked hairs. It grows in deep woods below 9,000 feet in the San Bernardino Mountains, Sierra Nevada, Cascade Range, and Warner Mountains and in northwestern California from Marin County northward and to British Columbia and the Rocky Mountains. It can be found in flower from June to August.

The carrot family (Apiaceae) consists of a large group of aromatic herbs with hollow stems, often decompound leaves, and

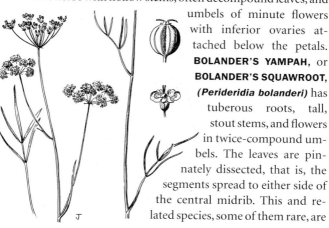

umbels of minute flowers with inferior ovaries attached below the petals. **BOLANDER'S YAMPAH,** or **BOLANDER'S SQUAWROOT,** *(Perideridia bolanderi)* has tuberous roots, tall, stout stems, and flowers in twice-compound umbels. The leaves are pinnately dissected, that is, the segments spread to either side of the central midrib. This and related species, some of them rare, are often conspicuous in drying meadows below 11,000 feet from San Diego County to British Columbia. It blooms from June to July. The tubers are edible and were a traditional Native American food.

MOUNTAIN SWEET-CICELY *(Osmorhiza chilensis)* has the aromatic quality of the other members of the carrot family. Perennial, slender-stemmed, and one to three feet high, it has rounded leaf blades divided into rather broad leaflets. The minute flowers are white, and the fruit is long and cylindrical. It grows in woods below 9,000 feet from San Diego County to Alaska and the Atlantic Coast, as well as in South America. It can be seen blooming from April to July.

LARGE-FRUITED LOMATIUM, or **LARGE-FRUITED HOG-FENNEL,** *(Lomatium macrocarpum)* is another perennial in the carrot family and 10 to 15 inches high. Its leaves are in a subbasal tuft and much divided into linear segments. The numerous, often hairy, flowers are white to yellowish or even purplish, and the flat fruits are one-half to two-thirds of an inch long. The species is found in dry, rocky places below 10,000 feet in the South Coast Ranges, Tehachapi Mountains,

Sierra Nevada, San Francisco Bay Area, Cascade Range, Klamath Mountains, and North Coast Ranges and northward to British Columbia. Flowering is from April to June.

Another member of the carrot family is **BREWER'S ANGELICA (Angelica breweri),** a conspicuous plant three to four feet high with large lance-shaped leaflets making up the large leaves. The inflorescence is also large and consists of 20 to 50 unequal primary divisions, or rays.

The petals and ovaries are hairy and produce oblong to oval fruits. This angelica is found on open wooded slopes from 3,000 to 10,000 feet in the Sierra Nevada and Cascade Range. Another montane species is Sierra angelica *(A. lineariloba),* which has linear leaflets less than half-an-inch wide. It can be found in the central, southern, and eastern Sierra Nevada and the White and Inyo Mountains. Both species bloom from June to August.

A small, stemless alpine plant in the carrot family is **CLEMENS' MOUNTAIN-PARSLEY (Oreonana clementis),** which has flowers and fruit in globose clusters on stems only slightly higher than the tuft of dissected, gray-hairy leaves only an inch or so in length. This is a plant of dry, granitic gravel from 5,500 to 13,000 feet in the southern Sierra Nevada. Flowering is from late May to August.

Cow-parsnip

One of the larger California members of the carrot family is **COW-PARSNIP *(Heracleum lanatum)***. Although the roots are perennial, the above-ground parts of this plant are produced new each year and can grow to a height of as much as nine feet in only a few months. The broad leaves are as much as 20 inches across. The large flower head is a compound umbel of 15 to 30 rays, or stems, each bearing a smaller umbel of white flowers. Cow-parsnip grows in moist places below 8,500 feet in most of California and to Alaska and the Atlantic Coast. It flowers from April to July.

Another large member of the carrot family is **SWAMP WHITE HEADS**, or **RANGER'S BUTTONS *(Sphenosciadium capitellatum)***. It is two to five feet high with leaves divided into several narrow or somewhat wider leaflets. The flowers are in large umbels, with each ray, or stem, bearing white heads of tiny flowers. This species grows in swampy places from 3,000 to 10,400 feet in the mountain ranges of southwestern California, Sierra Nevada, Cascade Range, North Coast Ranges, and Great Basin areas of eastern California and to Oregon and Idaho. It flowers in July and August.

Swamp white heads,
or ranger's buttons

With over 40 species in California and forming a conspicuous portion of our woody vegetation are the manzanitas (*Arctostaphylos* spp.), in the heath family (Ericaceae), which are characterized by a shrubby habit, smooth red stems, and clusters of small white or pinkish urn-shaped flowers. **PINEMAT MANZANITA (*Arctostaphylos nevadensis*)** is sprawling or prostrate and has intricately branched stems and light green leaves. It inhabits moist places or dry, rocky slopes in woods from 3,000 to 10,000 feet in the Sierra Nevada, Cascade Range, Klamath Mountains, and North Coast Ranges and northward to

Pinemat manzanita

Washington. It blooms from May to July. A common, but taller, manzanita found through most of the Sierra Nevada and northern mountains is white-leaved manzanita *(A. viscida)*, distinguished by its gray green leaves and sticky, glandular bracts, flower pedicels, and berries. A subspecies with glandular leaves and very glandular inflorescence branches is Mariposa manzanita *(A. viscida* subsp. *mariposa)*, which is found in the Sierra Nevada from Amador to Kern Counties. Both white-leaved manzanita and Mariposa manzanita bloom from February through April.

Another member of the heath family is **WESTERN BLUE-BERRY *(Vaccinium uliginosum* subsp. *occidentale)*,** a low

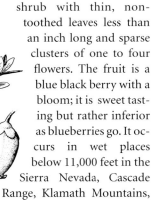

shrub with thin, non-toothed leaves less than an inch long and sparse clusters of one to four flowers. The fruit is a blue black berry with a bloom; it is sweet tasting but rather inferior as blueberries go. It occurs in wet places below 11,000 feet in the Sierra Nevada, Cascade Range, Klamath Mountains, and North Coast Ranges and to British Columbia and the Rocky Mountains. The flowers appear from June to July.

WESTERN AZALEA *(Rhododendron occidentale)*, also in the heath family, is a deciduous shrub one to 10 or more feet tall and has shredding bark. The large flowers are white to pinkish, and the upper lobes have a yellowish blotch, although there can sometimes be other color variations. This fragrant azalea grows in moist places below 7,500 feet in most of the mountain ranges of California and north to Oregon. It is a

Western azalea

conspicuous shrub on river banks in Yosemite Valley and blooms from April to August.

WESTERN LABRADOR-TEA (Ledum glandulosum) is a rather rigidly branched shrub from one to almost five feet tall with alternate, leathery leaves that have flat, sessile glands on the

Western Labrador-tea

underside, which give off a pleasant fragrance. In California, the inflorescences of many plants of this species have only a few flowers. The petals are white to cream yellow, measuring about one-fourth to one-third inch in length. Found in most of montane California, the Sierran form inhabits boggy and wet places from 4,000 to 12,000 feet and ranges to Trinity and Modoc Counties. The coastal form is usually found below 2,000 feet. The blooming period is from April to August.

WHITE-HEATHER (Cassiope mertensiana), another member of the heath family, is shrubby and has ascending branches four to 12 inches high and minute leaves keeled on the back.

White-heather

The white to pinkish bell-shaped flower is one-fourth inch long. It grows on rocky ledges and in crevices from 6,000 to 12,000 feet in the Sierra Nevada, Cascade Range, and Klamath Mountains and to Alaska and Montana. It can be found in bloom from July to August.

Also in the heath family is **PITYOPUS,** or **CALIFORNIA PINE-FOOT,** *(Pityopus californicus),* a waxy-white saprophyte that lacks chlorophyll and lives off dead matter in the soil. It is two to eight inches high and has scalelike leaves. Its white flowers are usually in a dense, terminal spikelike cluster, but they can also be solitary. The four to five petals are glabrous outside but densely hairy within. An uncommon plant that is not often encountered, it occurs in deep shade below 6,000 feet at scattered stations in the central and North Coast Ranges, Klamath Mountains, and southern Sierra Nevada and also in Oregon. It flowers from May to July.

WHITE-VEINED WINTERGREEN (Pyrola picta), also of the heath family, is a curious little plant growing from a branched rootstock. Its rubbery leaves are mottled or veined with white, and its blades are one to three inches long. The flowers are on a leafless stem four to eight inches tall and have cream to greenish petals. Some forms are almost leafless, and others have little or no whitening along the veins. The plants grow in humus in dry forests from 1,000 to 8,000 feet in most California mountains except in the desert, and they range north to British Columbia. The blooming period is from June to August. (See "Reddish Flowers" for a pink *Pyrola* species.)

The dogbane family (Apocynaceae) is well known in cultivation for plants such as oleander, periwinkle, and natal plum. A native species in our mountains is **BITTER DOGBANE (Apocynum androsaemifolium),** a

Bitter dogbane

Green gentian, or monument plant

smooth, diffusely branched perennial with drooping leaves paler beneath than above. The bell-shaped flowers have white lobes with pinkish veins. The fruit usually consists of two ripened pods (or follicles) that are pendulous at maturity and two to four inches long. This dogbane occurs in dry places below 8,000 feet in the San Bernardino and San Jacinto Mountains northward. Flowering is from June to August.

In the gentian family (Gentianaceae), most of our conspicuous montane plants have blue flowers, but **GREEN GENTIAN,** or **MONUMENT PLANT,** *(Swertia radiata)* has an open, greenish white corolla dotted with purple and measuring an inch or more across. Green gentian is a coarse plant and three to six feet high and has whorls of three to seven leaves and four-parted flowers. It is found in dry to damp places from 5,000 to 9,800 feet in the Sierra Nevada, Warner and Klamath Mountains, and North Coast Ranges and to Washington and the Rocky Mountains. Other related green gentians occur in most parts of the state. It flowers from July to August. (See "Bluish Flowers" for another *Swertia* species.)

To many Californians, the morning glory is exemplified by a weedy pest, the introduced orchard bindweed *(Convolvulus arvensis)*, but some of the native species in the morning glory family (Convolvulaceae) are more attractive. **SIERRA MORN-ING GLORY *(Calystegia malacophylla)*** is a low vine with grayish-woolly leaves, short trailing stems, and flowers an

inch or more long. Two subspecies occur in California. In the northern part of the state is *Calystegia malacophylla* subsp. *malacophylla* with widely triangular to more or less reniform leaves and brown to golden brown hairs on the plant. It is found from about 3,000 to 8,000 feet in the Sierra Nevada and Cascade Range. In the southern ranges is found *Calystegia malacophylla* subsp. *pedicellata,* with more narrowly triangular leaves and grayish hairs. It ranges from 1,000 to 6,500 feet in the South Coast Ranges, western Transverse Ranges, and San Francisco Bay Area. It is found on dry steep slopes and in chaparral and blooms from June to August.

An alpine plant in the phlox family (Polemoniaceae) is **WHITE-GLOBE-GILIA *(Ipomopsis congesta* subsp. *montana),*** a matted perennial with palmately lobed leaves, that is, the lobes spread out like the fingers from the palm of a hand. The small white flowers are crowded into heads and have the five-lobed corolla and three-parted style characteristic of the phlox family. This plant is found mostly in dry places from 7,000 to 12,000 feet in the Great Basin areas of eastern California. It blooms in June and July. (See "Reddish Flowers" for another *Ipomopsis* species.)

Another member of the phlox family is **HARKNESS' LINAN-THUS *(Linanthus harknessii),*** an annual with slender stems to about a foot high. The paired leaves are palmately divided into three to five linear lobes, and the filiform, or threadlike, ultimate branchlets of the stems bear small, whitish flowers. This plant grows in open sandy and gravelly places from 3,000 to 10,500 feet in the Sierra Nevada, Cascade Range, North Coast Ranges, and Great Basin areas of eastern California and to Washington. Other related species varying in technical characteristics also occur in various parts of California. This species flowers from June to August. (See "Reddish Flowers" for a pink *Linanthus* species.)

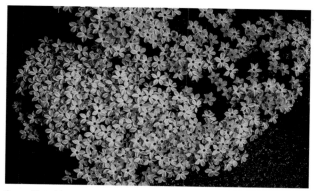

Spreading phlox

SPREADING PHLOX *(Phlox diffusa)* is a more or less prostrate perennial and somewhat woody at the base and has almost needlelike leaves about half-an-inch long. The flowers are mostly solitary at the ends of short, leafy branches and about half-an-inch wide. This plant grows on dry slopes and flats from 3,300 to 12,000 feet in most of montane California except the desert and southwestern part of the state. Related white-flowered species may be more cushionlike, have shorter leaves, and be more glandular. Spreading phlox blooms from May to August. (See "Reddish Flowers" for a pink *Phlox* species.)

In the waterleaf family (Hydrophyllaceae) is **IMBRICATE PHACELIA (Phacelia imbricata),** a rather coarse, often hispid (rough hairy) perennial that grows from a woody caudex and has stems one to two feet long. The lower leaves are lobed,

Imbricate phacelia

and the flowers are crowded in dense, coiled branches arranged in open clusters. The species is extremely variable and, in one form or another, it occurs on dry, often rocky places below 8,500 feet through much of California. It can be found blooming from April to June. Related species are found up to the timberline and beyond.

CALIFORNIA WATERLEAF (Hydrophyllum occidentale) has an elongate rhizome and hairy stems up to two feet high, although it is often much shorter. The bell-shaped corolla is white to lavender and about one-third inch long. It occurs on dryish or moist, somewhat shaded slopes from 2,000 to 10,000 feet in the Sierra Nevada, Cascade Range, San Francisco Bay Area, and mountains of northwestern California and then to Oregon and Idaho. Flowers appear from May to July.

Also in the waterleaf family is **DWARF HESPEROCHIRON (Hesperochiron pumilus),** a stemless perennial with leaves to about two inches long. The

California waterleaf

flowers are flat and open, measuring up to almost an inch across, and densely long hairy within. This plant occurs in moist, sometimes subalkaline places from 1,200 to 10,000 feet in the western Transverse Ranges, Tehachapi Mountains, Sierra Nevada, Cascade Range, Klamath Mountains, North Coast Ranges, and Great Basin areas of eastern California and

Dwarf hesperochiron

then to Washington. California hesperochiron *(H. californicus)*, which occurs from the San Bernardino Mountains northward, has a more funnelform corolla and is less hairy within. The flowering period for both species ranges from April to July.

FIVESPOT *(Nemophila maculata)*, also in the waterleaf family, is a lovely little plant with white flowers and a large, deep purple spot at the top of each petal. The corolla is rotate to bowl shaped and can be up to two inches across. The calyx lobes alternate with tiny, reflexed sepal-like appendages that are characteristic of most plants in the waterleaf family that have solitary flowers rather than coiled branches. Fivespot is closely related to the more well known and widespread baby blue-eyes *(N. menziesii)* and can be found in more or less moist, grassy or wooded slopes below 7,500 feet in the Sierra Nevada, as well as in the Sacramento Valley. It flowers from April to July.

Fivespot

The borage family (Boraginaceae) also has coiling flower clusters, or cymes, and its ovaries usually form four one-seeded nutlets. **ALPINE FORGET-ME-NOT *(Cryptantha humilis)*** is a cespitose, or denseley tufted, perennial with one to several

Alpine forget-me-not

very leafy stems arising from a woody caudex. The white flower is almost half-an-inch wide. The species is characteristic of high, dry ridges from 5,500 to 12,000 feet in the Great Basin areas of eastern California. Flowering is from June to August. A similar species with smooth instead of tubercled seeds is Sierra oreocarya (*C. nubigena*), found mostly on the eastern side of the Sierra Nevada and in the Inyo, White, and northern Desert Mountains.

TORREY'S CRYPTANTHA, or **WHITE FORGET-ME-NOT,** **(*Cryptantha torreyana*)** is the common annual species of the pine belt. It is usually rough hairy with slender, often openly branched stems four to 16 inches tall. The flowers are small and usually in a terminal inflorescence. This species grows on dry, more

Torrey's cryptantha, or white forget-me-not

or less open slopes from 1,500 to 6,500 feet from Santa Cruz and Kern Counties northward to British Columbia.

SHORT-FLOWERED MONKEYFLOWER (Keck-iella breviflora), in the figwort family (Scrophulariaceae), is a slender, shrubby plant, is usually much branched from below, and has opposite, almost stemless leaves that are three veined. The glabrous to sticky hairy inflorescence consists of white to cream flowers about an inch long with purple or pink stripes on the petals. The corolla throat has two ventral ridges with short, knobbed hairs and faint brown spots. It is found on rocky slopes in forest and chaparral below 12,000 feet in the Sierra Nevada, North and South Coast Ranges, and western Transverse Ranges, as well as in Nevada. It flowers from May to June.

Dwarf chamaesaracha

In the nightshade family (Solanaceae) are such familiar plants as potato, tobacco, pepper, and tomato, and most members of this family have quite a strong odor. **DWARF CHAMAE-SARACHA (Chamaesaracha nana)** is a low perennial growing from tough, slender underground rootstocks and has entire leaves one to two inches long. The corolla is white with five basal green spots, and the fruit is a berry that is dull white to yellowish. This plant occurs on sandy flats from 5,000 to 9,000 feet in the Sierra Nevada, Cascade Range, and Great Basin areas of eastern California. It can be found flowering from May to July.

BUSH CHINQUAPIN (Chrysolepis sempervirens), along with the oaks, belongs to the beech family (Fagaceae). It is a rounded shrub two to eight feet high and has oblong, almost toothless leaves one to three inches long that are golden or rusty woolly beneath. The staminate flowers, appearing in July and August, are very ill smelling and arranged in long catkins (long clusters of petalless, unisexual flowers). The pistillate flowers produce spiny burs enclosing one to three nuts that take two years to mature. The species occurs in dry, rocky places from 2,500 to 11,000 feet from the San Jacinto Mountains to Oregon.

Bush chinquapin

Mountain dogwood

MOUNTAIN DOGWOOD *(Cornus nuttallii)* is a deciduous ar-
borescent shrub or small tree of the dogwood family (Cor-
naceae). The leaf blades are commonly two to four inches
long, and the actual flowers are in a small, inconspicuous
head surrounded by large white, or sometimes pinkish, persistent bracts that are quite petal-like and showy. These bracts

usually appear in spring before the leaves and give this species of dogwood its lovely appearance. It is found in mountain woods below 6,500 feet. It occurs only locally in the mountains from San Diego to Los Angeles Counties but is more widespread through central and northern California and northward to British Columbia. It is a popular spring attraction in Yosemite Valley and blooms between April and July.

BLUE ELDERBERRY (*Sambucus mexicana*), of the honeysuckle family (Caprifoliaceae), is a low shrub at higher altitudes and a taller one at lower elevations. It has opposite, compound

leaves divided into large, toothed leaflets. The very numerous, small, white flowers are in flat-topped clusters and form small, bluish berries about one-fourth inch in diameter. Blue elderberry occurs in open places below 10,000 feet from San Diego County northward through most of California and to British Columbia and Alberta. Its blooming period is from March to September.

Blue elderberry

RED ELDERBERRY (*Sambucus racemosa* var. *microbotrys*) is a low shrub with a rank odor and cream-colored flowers in dome-shaped or sometimes more elongate clusters. The red fruits are about one-sixth inch long and lack the powdery bloom of the berries of blue elderberry (*S. mexicana*). Red elderberry is common in moist places from 6,000 to 11,000 feet in the San Bernardino Mountains, Sierra Nevada, Cascade Range, and mountains of northwest California and eastward to the Rocky Mountains. It blooms from June to August.

Red elderberry

CALIFORNIA VALERIAN *(Valeriana californica)* is in the valerian family (Valerianaceae), which is characterized by small flowers that are often pouched or spurred on one side. This species of valerian is a perennial with strong-scented underground parts and paired leaves that may be entire or divided. The calyx is split below, especially as it matures, into 12 to 17 bristles. The species is found in moist, or sometimes dryer, places from 5,000 to 11,000 feet in the Sierra Nevada and most northern California mountain ranges and in Oregon. It flowers from July to September.

The sunflower family (Asteraceae), with its flowers minute and crowded into heads surrounded below by leaflike bracts, or phyllaries, is prominent in California. In the genus *Chaenactis,* the flowers are all small, tubular disk flowers, the outer larger than the inner. **SANTOLINA CHAENACTIS,** or **SANTO-**

California valerian

LINA PINCUSHION FLOWER, *(Chaenactis santolinoides)* is a perennial with simple or branched stems up to about a foot high. The leaves are mostly in basal rosettes, and the cream or white flower heads are about half-an-inch high. It is found in open woods and on dry ridges from 4,500 to 9,000 feet in the Sierra Nevada and Transverse Ranges. It flowers from June to July.

Related species of *Chaenactis* are widespread in California, such as **DUSTY MAIDENS *(Chaenactis douglasii)*,** which is more or less floccose, or woolly, when young. The loosely-woolly leaves are one to six inches long and occur along the stem, as well as in a basal rosette. The white to pinkish flower heads are on stems less than four inches tall. Varied forms of this species range from 3,000 to 11,500 feet in the Sierra Nevada, Cascade Range, mountains of northwestern California, northern Desert Mountains, and Great Basin areas of eastern California and northward to British Columbia. It can be found blooming in June and July.

YARROW, or **MILFOIL,** *(Achillea millefolium)* is another member of the sunflower family. A perennial, quite aromatic herb with very finely dissected leaves, it has numerous small heads of short, rounded ray flowers that surround several central disk flowers. It is a circumboreal (worldwide) species and is found in many habitats up to 11,500 feet through most of California. Flowers appear between March and August.

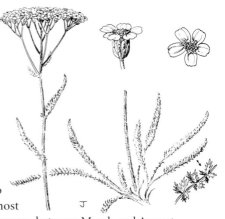

TRAIL PLANT *(Adenocaulon bicolor),* a perennial growing from slender rootstocks, is white woolly below and glandular in the

Trail plant

upper parts. The leaves are near the base and white woolly on the underside, giving this species its common name—if you turn the leaf over as you walk, you can supposedly find your way back by following the trail of white leaves. The white, tubular

disk flowers are in heads with an involucre below that consists of a single row of subequal green phyllaries. There are no ray flowers. Trail plant is found in moist, shaded woods below 6,500 feet in the Sierra Nevada, Cascade Range, and Coast Ranges from Santa Cruz County northward and to British Columbia and Michigan. It flowers from June to August.

ROSY EVERLASTING, or **ROSY PUSSYTOES, (Antennaria rosea)** forms leafy, gray woolly mats from which arise flowering stems that can be from only a few inches to almost a foot in height. The staminate and pistillate flower heads are on separate plants, and the minute flowers are in woolly heads with dry, scalelike phyllaries that may be white or almost rose in color. It is found in dryish to moist, more or less wooded places from 4,000 to 12,000 feet in the San Jacinto, San Bernardino, and San Gabriel Mountains, Sierra Nevada, Cascade Range, Klamath Mountains, and Great Basin areas of eastern California and northward to Alaska and Ontario. Other species may have greenish leaves, and some can have phyllaries that are brownish to a dirty green. The flowering period is from June to August.

Pearly everlasting

PEARLY EVERLASTING (Ana-phalis margaritacea), also of the sunflower family, is a white-woolly perennial with slender, running rootstocks. The leafy, erect stems are unbranched and have stemless leaves that are mostly one to four inches long and often greener on the upperside than beneath. The flower heads are pearly white. It grows in wooded areas and along roadsides below 10,500 feet in the San Bernardino Mountains, central and North Coast Ranges, Sierra Nevada, Cascade Range, and Modoc Plateau and also to Alaska, the Atlantic Coast, and Eurasia. It blooms from June to August.

Western eupatorium

WESTERN EUPATORIUM *(Ageratina occidentalis)* is another member of the sunflower family that does not have ray flowers. A more or less woody tufted perennial, it has mostly alternate leaves with rather compact clusters of flower heads at the ends of the branches. The tubular disk flowers are white, pink, or red purple. Found about rocks from 6,500 to 12,000 feet, it occurs in the Sierra Nevada, Cascade Range, and Warner Mountains and in most of northwestern California northward to Washington and eastward to Utah. Flowering is from July to September.

WHITE-FLOWERED HAWKWEED *(Hieracium albiflorum)*, in contrast to western eupatorium *(Ageratina occidentalis)*, is composed only of ray flowers. It is an erect perennial and densely hairy in its lower parts. The mostly basal leaves are up to six inches long, but they are reduced on the upper stem.

The white-flowered heads are in a loose, branched inflorescence, and the involucres, formed by the overlapping leaflike phyllaries below the head, are almost half-an-inch long. This plant occurs on dry, open, wooded slopes below 9,700 feet from San Diego County to Alaska and Colorado. It blooms from June to August. (See "Yellowish Flowers" for another *Hieracium* species.)

CAMAS *(Camassia quamash)*, a member of the lily family (Liliaceae), grows from a perennial bulb and has a leafless stem one to three feet tall with a basal whorl of linear leaves. The flowers are a deep blue violet to bright blue and over an inch wide. Highly variable, it is theorized that the many localized forms may be the result of the bulbs having been a popular food of the Native Americans and having often been traded among tribes. Sheets of this lovely plant can often be seen in mountain meadows up to 11,000 feet as far south as Marin and Tulare Counties and as far north as British Columbia. Camas blooms from May to August.

Camas

WILD-HYACINTH *(Dichelostemma multiflorum)* is a handsome member of the lily family, formerly in the *Brodiaea* genus. The numerous blue flowers are in a tight, rounded umbel at the top of a flower stem that can be up to three feet tall. Each flower is shaped like an upside-down bell with a narrow tube and widely spreading corolla lobes above. The three to four narrow basal leaves are usually withered by the time the flowers appear. This plant occurs in many habitats up to 7,000 feet in the northern Sierra

Nevada, in most of northwestern California, and in Oregon. Twining brodiaea, or snake-lily, *(D. volubile)* is another striking member of this genus and is often encountered in the northern mountains. It is similar to wild-hyacinth, but the flower tube is shorter and wider, and the flowers are pink but with a blue tinge. The stem, which can be up to five feet long, is weak and twines around nearby vegetation for its support, thus giving the plant its common names. Both species bloom in May and June.

BEAVERTAIL-GRASS *(Calochortus coeruleus)*, also in the lily family, is a low plant that has a basal leaf up to eight inches long and usually one to 10 flowers. The broad, bluish petals have a smooth inner surface except for the beard above the nectar gland and the fringe on their margins. This plant grows in open, gravelly places in woods from 2,000 to 8,000 feet in northwestern California, the Cascade Range, and the Sierra Nevada. Several related species varying in color and technical characteristics occur in northern California. Beavertail-grass blooms from May to July. (See "Whitish Flowers" for another *Calochortus* species.)

Beavertail-grass

Western blue flag

In the iris family (Iridaceae), **WESTERN BLUE FLAG** *(Iris missouriensis)* grows from a stout rhizome and has a typical iris flower structure, pale lilac sepals, and even paler petals with lilac purple veins. It grows largely between 3,000 and 11,000 feet and is considered a noxious weed by many farmers and ranchers because its bitter leaves are avoided by most range animals. It blooms in May and June. The iris genus (*Iris* spp.) is one of the most widespread in western North America; it is found in meadows and moist flats and ranges from the mountains of San Diego County to British Columbia, South Dakota, and Coahuila, Mexico. (See "Yellowish Flowers" for another *Iris* species.)

Larkspurs (*Delphinium* spp.) are one of the largest groups in the buttercup family (Ranunculaceae), and **MOUNTAIN LARKSPUR** *(Delphinium glaucum)* is one of the physically largest western species in this group. It is a coarse-stemmed leafy plant usually about five feet tall with leaves three to seven inches across and divided into broad segments that may be toothed or deeply incised. The inflorescences are four to 12

Mountain larkspur

inches long and usually have 50 or more light to dark violet purple flowers with spurs that can be an inch long. Found in wet meadows and near streams from 5,000 to 12,000 feet, this species occurs in the San Bernardino and San Gabriel Mountains, Sierra Nevada, and Klamath Mountains, thence to Alaska and the Rocky Mountains. It can be found blooming from July to September.

HIGH MOUNTAIN LARK-SPUR *(Delphinium polycladon)* has several unbranched stems, mostly two to four feet high, and basal palmatifid leaves, that is, leaves that are very deeply divided into lobes spreading like the fingers from the palm of a hand. The inflorescence is often one sided with three to 35 dark blue to blue purple flowers, with sepals about half-an-inch long and spurs from one-half to one inch long. It is found among rocks and willows along creeks and in meadows from 7,500 to 12,000 feet in the Sierra Nevada and White and Inyo Mountains. Flowering is between July and September.

Another widespread species of the buttercup family is **WESTERN MONKSHOOD** *(Aconitum columbianum),* a highly variable complex with many local forms. With rather tuberous roots, it has mostly erect, stout stems from two to over six feet tall and well-distributed, palmately lobed leaves. The flowers, over half-an-inch long, are in a rather loose inflorescence and are usually purplish blue, with the upper sepal arched into a hood. Western monkshood grows in moist places such as willow thickets from 2,000 to 11,500 feet from the Sierra

Western monkshood

Nevada through most of the northern California mountain ranges and northward to British Columbia. It blooms from July to August.

In the pea family (Fabaceae) are the lupines (*Lupinus* spp.), one of California's prides. **BREWER'S LUPINE** *(Lupinus breweri)* is a low, matted perennial with silvery-silky foliage and stems less than eight inches high. The compound leaves are crowded and consist of five to 10 leaflets less than an inch long. The small violet flowers are less than half-an-inch long.

Brewer's lupine

This species has three varieties that can be found on dry, rocky slopes and benches between 3,000 and 13,000 feet from the San Bernardino Mountains to Oregon, blooming from June to August.

GRAPE SODA LUPINE _(Lupinus excubitus)_ has five varieties that range from quite woody at the base to almost herbaceous. They are rather silvery silky, prostrate to erect, and have seven to 10 leaflets an inch or more long. The rather fragrant flowers are blue to violet and about half-an-inch long. It is common on dry slopes and in rocky places below 10,000 feet from the southern Sierra Nevada and Tehachapi Mountains

Grape soda lupine

Large-leaved lupine

to Baja California and also in the desert and eastern Sierra Nevada. The flowering period is from April to June.

A very showy lupine is **LARGE-LEAVED LUPINE** *(Lupinus poly-phyllus var. burkei)*. Up to five feet high, the stout stem has leaves with five to 17 leaflets from two to six inches long. The numerous blue to purplish flowers are about half-an-inch long. Growing in wet places, this variety can be found at elevations from 5,000 to 10,000 feet in the San Jacinto Mountains northward to the Sierra Nevada and also into Oregon, Nevada, and Idaho. Other varieties of this species grow in the Coast

Ranges and mountains of northern California. Flowers appear between May and August.

Another member of the pea family is **FEW-FLOWERED PEA**, or **BUSH PEA**, *(Lathyrus brownii)*, an erect perennial with angled stems one to two feet long and with two to six orchid to violet purple flowers that become bluish in age. The compound leaves have six to 10 leaflets and simple or forked tendrils. This pea is found on dry slopes from 3,000 to 6,000 feet in the north and central Sierra Nevada and in the northern mountain ranges of California into Oregon. It flowers from April to June.

Western dog violet

The violet family (Violaceae) has many representatives in California. One of the mountain species is **WESTERN DOG VIOLET** *(Viola adunca)*, which has deep to pale violet flowers and a spur one-fourth to one-half inch long. It has a branched stem with basal and cauline leaves and spreads by means of slender rootstocks. It can be found on damp banks and at edges of meadows up to 11,500 feet throughout most of montane California to Alaska and Quebec. It flowers between March and July. (See "Yellowish Flowers" for another *Viola* species.)

Western blue flax

WESTERN BLUE FLAX *(Linum lewisii)* is a glabrous perennial in the flax family (Linaceae) and usually has several leafy stems up to two-and-a-half feet high with numerous narrow leaves. A true flax, it has very fibrous and tough tissues in the stem. The blue flowers have five blue, sometimes white, petals about half-an-inch long, which fall away easily when the plant is disturbed. It grows on dry slopes and ridges below 11,000 feet from Baja California to Alaska and eastern Canada and Texas. It blooms between May and September.

MAHALA MAT, or **SQUAW CARPET,** *(Ceanothus prostratus)* of the buckthorn family (Rhamnaceae), is a prostrate shrub with its branches rooting and forming mats a yard or more across. The opposite leaves are light green and often have three to nine sharp teeth near the tip. The light to deep blue

Mahala mat, or squaw carpet

flowers are in small clusters. It inhabits open flats in pine forests from 2,000 to 7,000 feet from the central Sierra Nevada through most northern California mountain ranges and then northward to Washington. It flowers from April to June.

I always associate gentians (*Gentiana* spp.) with mountain meadows, as far as California is concerned. One of our more conspicuous species is **EXPLORER'S GENTIAN (*Gentiana calycosa*),** in the gentian family (Gentianaceae). It is a perennial that grows from a root crown and has thick, cordlike roots. Each of the several stems bears many broad leaves and deep blue bell-shaped flowers an inch or more long

Explorer's gentian

with five corolla lobes. Found in meadows, on stream banks, and in other wet places from 4,000 feet in the Klamath Mountains and Cascade Range up to 13,000 feet in the Sierra Nevada, this gentian grows as far north as British Columbia and also to Montana. It blooms from July to September.

ALPINE GENTIAN (*Gentiana newberryi*) is a decumbent perennial with its stems arising from below a rosette of broadly spatulate leaves about one to three inches long. There can be one to five flowers per stem, and the broadly funnelform corolla is white to deep blue, often with dark purplish bands on the outside. It occurs in moist meadows and on banks, mostly from 4,000 to 13,000 feet, in the Sierra Nevada, Klamath Mountains, Cascade Range, and White and Inyo Mountains and to southern Oregon. Flowering is from July to September.

Alpine gentian

Sierra gentian

SIERRA GENTIAN *(Gentianopsis holopetala)* is an annual with small, slender roots and several stems, each having only one flower. The blue, funnelform corolla is one to two inches long and four lobed. It grows in wet meadows from 6,000 to 13,000 feet in the central Sierra Nevada and White and Inyo Mountains. Another montane annual is hiker's gentian *(G. simplex)*, which has a single stem that ends in a solitary flower with toothed or fringed lobes. It is more widespread, occurring in most of northern California, the Sierra Nevada, and the San Bernardino Mountains. Both species flower from July to September.

Another gentian is **FELWORT** *(Swertia perennis)*, a perennial with a short rootstock and a single, erect, unbranched stem up to almost two feet tall. The leaves are mostly two to six inches long in a basal rosette, and the few, short cauline leaves can be opposite or alternate. The flowers, in a simple or branched terminal cluster, are four or five lobed, about half-an-inch long, and bluish white to violet blue with darker veins. It is found in meadows, bogs, and damp places from 7,500 to 10,500 feet in the southern Sierra Nevada, from Oregon to Alaska, and in the Rocky Mountains and Eurasia. It flowers from July to September. (See "Whitish Flowers" for another *Swertia* species.)

In the phlox family (Polemoniaceae), **JACOB'S LADDER (Polemonium occidentale)** is a perennial growing from a rootstock and has solitary erect stems one to over three feet tall. The lower leaves have 13 to 17 leaflets, and the upper leaves have nine to 13. The blue flowers usually have a yellow throat and are spreading bell shaped and about an inch across. It is found in wet places from 3,000 to 11,000 feet in the San Bernardino Mountains, Sierra Nevada, and Klamath Mountains, thence to Alaska and the Rocky Mountains. It blooms from July to August.

Jacob's ladder

Sky pilot

SKY PILOT *(Polemonium eximium)* is a viscid perennial four to 16 inches high with a woody base and a strong, musky odor. The leaves are divided into numerous three- to five-parted leaflets. The flowers are crowded into a head and are narrow funnelform to cylindrical, about half-an-inch long, and mostly deep blue in color. The species grows on dry, rocky ridges and slopes from 10,000 to 14,000 feet in the central and southern Sierra Nevada. Flowering is in July and August.

In the waterleaf family (Hydrophyllaceae), the phacelias (*Phacelia* spp.) have many forms. A low, annual type is represented by **WASHOE PHACELIA *(Phacelia curvipes),*** which is branched and diffuse and up to about six inches high. The leaves are largely basal and entire, although in some related species they are few lobed. The broadly bell-shaped corolla is almost one-third inch long and blue to violet with a white throat. It is a species found on dry slopes from 2,000 to 9,000 feet in the Transverse Ranges, Tehachapi Mountains, southern and eastern Sierra Nevada, and Mojave Desert and then to Utah and Arizona. It can be found blooming between April and June.

Washoe phacelia

WATERLEAF PHACELIA *(Phacelia hydrophylloides)* has a thick, woody taproot and several spreading to ascending stems that can be almost a foot long. The leaves are generally one to two inches long and range from coarsely toothed to deeply lobed. The flowers are densely clustered in a head, violet blue to white, broadly bell shaped, and half-an-inch across. It is occasional in dry woods from 5,000 to 10,000 feet from Tulare County in the Sierra Nevada through the Cascade Range to

Waterleaf phacelia

Draperia

Oregon and Nevada. The blooming period is from June to August. (See "Whitish Flowers" for another *Phacelia* species.)

Another member of the waterleaf family is **DRAPERIA** *(Draperia systyla)*, a low, diffuse perennial with slender stems and a branched root crown. The leaves are in pairs and one to two inches long. The corolla is pale violet to white or pink and about half-an-inch long. Draperia grows on dry slopes in woods below 10,000 feet in the Sierra Nevada, Klamath Mountains, and Cascade Range. It is named for J.W. Draper, an American historian. The flowers appear between May and August.

Streamside bluebells, or lungwort

STREAMSIDE BLUEBELLS, or **LUNGWORT,** *(Mertensia ciliata)* is a member of the borage family (Boraginaceae). It is a rather coarse perennial, about two to five feet tall, and has many bright green leaves. The inflorescence is open in age and has numerous, rather tubular, light blue flowers half-an-inch or more long. It is found in moist places, usually in the shade, from 5,500 to 12,000 feet in the northern Sierra Nevada, White and Inyo Mountains, and Modoc Plateau and into Oregon and Nevada. It blooms from May to August.

Likewise in the borage family, but with flowers more like the familiar forget-me-not *(Myosotis latifolia),* is **JESSICA'S STICKSEED *(Hackelia micrantha)*.** It has erect or ascending stems that are one to four feet high and grow from a heavy, perennial root. The leaves are well distributed along the stems, which end in an open inflorescence of several divergent coiled branches with pale blue flowers about one-sixth inch wide. The nutlets are prickly, giving the plants in this genus their common name. This stickseed is common in moist

places from 2,000 to 11,000 feet from the Sierra Nevada through most of the northern California mountain ranges and to as far as British Columbia. It blooms in July and August.

Gray-leaved skullcap

GRAY-LEAVED SKULLCAP (*Scutellaria siphocampyloides*) is in the aromatic mint family (Lamiaceae), having paired leaves, a two-lipped corolla, and fruit composed of four one-seeded nutlets. Members of this genus have a crestlike projection on the back of the calyx, hence, their common name. This species has deep violet blue flowers an inch long with glandular hairs. The stems have minute curled or appressed hairs, and the upper leaves have entire margins. It grows in gravelly or rocky places below 7,500 feet throughout most of California. Other species differ in size, color, and placement of flowers, as well as leaf shape and type of pubescence (hairiness). Flowering is from May to July.

Our native **SELF-HEAL** *(Prunella vulgaris var. lanceolata)* is also in the mint family. A low, perennial herb, it has leaves one to three inches long and stems four to 20 inches high. The two-lipped flowers are densely clustered, violet to purplish, and measure half-an-inch or longer in length. It is found in moist woods and near ditches below 8,000 feet in most of montane California and also ranges to

Self-heal

Alaska and the Atlantic states. It can be seen blooming between May and September.

Sages are also in the mint family, and **THICK-LEAVED SAGE,** or **ROSE SAGE,** *(Salvia pachyphylla)* is the most common mountain species of sage in southern California. It is a somewhat sprawling subshrub and has grayish leaves one to two inches long and dense spikes of flowers with purplish bracts below. The large blue to violet blue corolla is almost an inch long, narrow, and two lipped and has very long stamens extending well beyond the flower. This sage frequents dry, rocky places between 5,000 and 8,000 feet from the southern Sierra Nevada and Desert Mountains through the Tehachapi and

Thick-leaved sage, or rose sage

Sierran penstemon, or Sierran beardtongue

San Bernardino Mountains and the Peninsular Ranges to Baja California. It flowers from July to September.

SIERRAN PENSTEMON, or **SIERRAN BEARDTONGUE, (_Penstemon heterodoxus_)** is in the figwort family (Scrophulariaceae). It is a slender-stemmed perennial from two inches to two feet tall with mostly basal leaves. The glandular inflorescence has two to four rather distinct, many-flowered clusters. The blue purple corolla is narrow and about half-an-inch long and has equal upper and lower lips. It is found on rocky slopes and in alpine meadows from 3,000 to 12,000 feet in the Sierra Nevada, Cascade Range, and Klamath, White, and Inyo Mountains. It blooms in July and August. (See "Reddish Flowers" for another *Penstemon* species.)

TORREY'S FEW-EYED MARY (*Collinsia torreyi*), also of the figwort family, is in a genus that is remarkable for having a lower corolla lip with a boatlike middle lobe, or keel, that encloses the style and stamens—very much like in the pea family (Fabaceae), but in the figwort family, it is attached to the side lobes rather than separate from them. This species of *Collinsia* is erect, widely branched, and can grow to a height of about 10 inches. The corolla is about one-third to one-half inch long and has a broadly rounded basal pouch, a pale upper lip with purple dots, and a longer, deeper blue lower lip. It is found from 2,500 to 13,000 feet from the Transverse Ranges through the Sierra Nevada and northern California in several varieties, all blooming between May and August.

AMERICAN ALPINE SPEEDWELL (Veronica wormskjoldii), another member of the figwort family, is a perennial with erect

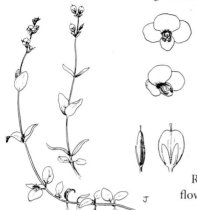

stems four to 16 inches
high, generally with four
to seven pairs of leaves on
each stem. The flowers are
terminal, with spreading,
bluish, four-lobed corol-
las one-fourth inch in diameter. It is found in wet
places, generally between
8,000 and 11,500 feet, in
the Sierra Nevada, Cascade
Range, and Modoc Plateau. It
flowers from June to August.

PURPLE NIGHTSHADE (Solanum xanti) is a shrub or subshrub
in the nightshade family (Solanaceae). Up
to three feet tall and more or less grayish
hairy, it has a spreading, deep violet to
dark lavender corolla up to an inch or
more in diameter and produces greenish,
round berries. It occurs in dry places,
such as along trails, up to 9,000 feet
through most of California except the
Cascade Range and Central Valley. Plants
at lower elevations are often woodier and
much taller than those at higher elevations.
The flowers appear between May and September.

In the bellflower family (Campanulaceae), **CALIFORNIA HARE-
BELL (Campanula prenanthoides)** is a perennial with slender
rootstocks and slender, reclining or erect stems that can be
from eight inches to as much as five feet tall. The flowers are
generally in scattered clusters with two to five bright blue flow-

California harebell

ers per cluster. The cylindric to bell-shaped corollas are almost half-an-inch long with extended styles that can be almost an inch long. It grows in dryish, wooded places below 6,500 feet in the Coast Ranges from Monterey County northward, the Sierra Nevada, the Cascade Range, and southern Oregon. Flowers appear from June to September.

PORTERELLA *(Porterella carnosula),* also in the bellflower family, is an erect annual, branched, and usually only a few inches tall. The corolla is blue with a yellow or whitish eye. It is strongly two lipped and one-fourth to one-third inch long. It occurs in wet places, often in masses, between 5,000 and 10,000 feet from the northern Sierra Nevada to the Cascade Range and in the Great Basin areas of eastern California. It is very similar to *Downingia* and *Lobelia* species of the same family but differs from the former because it has pedicelled

Porterella

flowers (each on a short stem) instead of being stemless and from the latter because is it an annual and has cylindrical or obovate fruit rather than spheric. Porterella flowers from June to August.

In the large sunflower family (Asteraceae), which has flowers arranged in heads of small tubular disk flowers or narrow,

Leafy aster

petal-like ray flowers, or both, **LEAFY ASTER (Aster foliaceus)** is an erect perennial that grows from a creeping rootstock and reaches a height of four inches to two feet. The leaves can be up to six inches long on the lower part of the stem but are smaller above. The petal-like ray flowers are purple to blue or violet and measure half-an-inch or longer. The numerous disk flowers are yellow. The flower head is surrounded below by several series of over-

Alpine aster

lapping leaflike phyllaries. Leafy aster occurs from 5,000 to 10,500 feet in damp places or on ridges in the San Bernardino and San Jacinto Mountains and Sierra Nevada and most northern California mountains. The flowering period is July and August.

ALPINE ASTER *(Aster alpigenus* var. *andersonii)* is a decumbent to erect perennial four to 16 inches tall and has basal, tufted grasslike leaves. The flower heads are solitary and showy, sometimes an inch or more in diameter. It grows in meadows from 5,000 to 12,000 feet in the San Jacinto Mountains, Sierra Nevada, White and Inyo Mountains, and most mountain ranges of northern California. It blooms from June to September.

Closely related to *Aster,* but with the phyllaries usually not overlapping in several series, is the genus *Erigeron.* **DWARF DAISY *(Erigeron pygmaeus)*** is low and tufted and grows from a woody taproot. It is more or less glandular and hairy throughout, its leaves are in a dense rosette, and the solitary

Dwarf daisy

flower heads are on erect, leafless stems only one to two inches high. The phyllaries are purplish or black purple, and the outer petal-like ray flowers are purple to lavender and about one-fourth inch long. The disk flowers in the center are yellow. This daisy occurs on rocky slopes and flats from 9,500 to 13,500 feet in the central and eastern Sierra Nevada. It flowers in July and August.

Another asterlike plant is **WANDERING DAISY (Erigeron peregrinus var. callianthemus)**, a fibrous-rooted perennial that can be up to almost two feet high and is leafy throughout. Its 30 to over 100 ray flowers are rose purple to white and about half-an-inch long, making it a very handsome daisy,

Wandering daisy

Hoary aster

as so many plants in this family are called. It grows in meadows from 4,000 to 11,000 feet in the Sierra Nevada, Cascade Range, and Klamath and Warner Mountains and northward to Alaska. It blooms between July and September.

In the same group of genera with the common name "aster" or "daisy," and much like *Aster*, but with a taproot instead of a rhizome or fibrous roots, is **HOARY ASTER** *(Machaeranthera canescens var. canescens)*. Its stems are six inches to almost two feet high, and the stem leaves are well developed. The one to many heads have eight to 15 narrow violet ray flowers, and

the leaflike phyllaries below the head are generally in five to 10 series and somewhat spreading reflexed. There are several other varieties, but this one occurs in various open montane habitats from 6,500 to 10,000 feet in the Sierra Nevada and Cascade Range northward to Washington and Canada and also in the Transverse Ranges of southern California. It flowers from July to September.

FALSE-ASPHODEL *(Tofieldia occidentalis),* of the lily family (Liliaceae), has a slender, perennial rootstock, stems one to three feet tall, and linear leaves two to eight inches long. The light yellow flowers are about one-sixth inch in diameter. This is a plant of boggy places and meadows below 10,000 feet in the Sierra Nevada and mountains of north-western California to southern Oregon. Flowers appear in July and August.

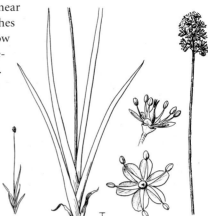

SLENDER TRITELEIA *(Triteleia montana),* also in the lily family, though sometimes placed in the onion family (Amaryllidaceae) and formerly in the *Brodiaea* genus, has more or less rough stems two to 10 inches high and leaves four to 12 inches long. The flowers are yellow, or purplish in age, and have brown midveins on the outside. This plant is locally rather plentiful on gravelly plains and granitic ridges from 4,000 to 9,800 feet in the northern and central Sierra Nevada and flowers in June and July.

Golden brodiaea, or pretty face

A similar *Triteleia* that is more widespread and has many sub-species is **GOLDEN BRODIAEA,** or **PRETTY FACE, (*Triteleia ixi-oides*).** It is similar to slender triteleia *(T. montana),* but the flowers have quite different-looking stamens: they are un-equal in length, have wide filaments, and have two pointed, forked appendages behind the anthers. This plant is often found in sandy soils in coniferous and mixed forests below 10,000 feet in the Sierra Nevada, central and South Coast Ranges, Klamath Mountains, and Cascade Range and into Oregon. It blooms between May and August. Two similar species grow in the southern ranges—Dudley's triteleia *(T. dudleyi),* in which the anther appendages are blunt rather than pointed, and calliprora *(T. lugens),* which has no anther appendages.

ALPINE LILY (*Lilium parvum*) has a horizontal, stemlike bulb and stems one to five feet tall or more. The leaves are light

Alpine lily

green, two to six inches long, and either scattered along the stem or in whorls. The orange to dark red flowers are spotted maroon, and the perianth segments are about one-and-a-half inches long or more. This lily inhabits boggy places at the edge of swamps or streams, often among alders or willows, from 4,500 to 9,500 feet throughout the northern and central Sierra Nevada. It blooms from July to September.

SIERRA LILY (*Lilium kelleyanum*) also has a horizontal, stem-like bulb. Its stems are two to six feet high, and its leaves about two to six inches long in whorls of two to five. The nodding, fragrant, yellow to yellow orange flowers are strongly recurved and have very long stamens with magenta to dull red anthers about one-fourth inch long. The perianth segments are about one to two inches long. This

Sierra lily

lily is found on wet banks and in boggy places from 7,000 to 10,000 feet in the central and southern Sierra Nevada. It flowers in July and August.

LEOPARD LILY, or **PANTHER LILY,** *(Lilium pardalinum)* is one of California's more conspicuous lilies. The stout stems are three to almost nine feet tall with one to eight whorls of nine to 15 leaves per whorl, plus some scattered ones as well. The nodding flowers have magenta, orange, or yellow anthers about half-an-inch long and orange to red perianth segments with maroon spots outlined in yellow or orange. They are recurved from the middle or below. It has several varieties and forms large colonies on stream banks and in springy places below 6,500 feet in most of montane California. It can be found flowering from May to July. (See "Whitish Flowers" for other *Lilium* species.)

Leopard lily, or panther lily

One of the few California irises that is not always blue is **HARTWEG'S IRIS *(Iris hartwegii)*,** of the iris family (Iridaceae). There are several subspecies, but most have rather narrow petals varying from deep yellow and lavender to pale yellow and cream and are one-and-a-half to over three inches long. They occur on wooded slopes from 2,000 to 7,500 feet, mostly in the Sierra Nevada and southern Cascade Range. A purple to bluish violet form, *I. hartwegii* subsp. *australis,* occurs largely in dry woods in the San Bernardino and San Jacinto Mountains and in the eastern San Gabriel Mountains. The blooming period is in May and June. (See "Bluish Flowers" for another *Iris* species.)

Hartweg's iris

Also in the iris family, with its leaves in two opposite ranks, is **DREW'S GOLDEN-EYED-GRASS,** or **DREW'S YELLOW-EYED-GRASS, *(Sisyrinchium elmeri)*,** a slender-stemmed plant less than eight inches tall, with narrow leaves less than half the length of the flower stems. The orange yellow flowers have dark brown veins and can be almost an inch in diameter. It is a plant of boggy and wet places between 4,000 and 8,500 feet in the San Bernardino Mountains, Sierra Nevada, Klamath Mountains, and southern Cascade Range. Flowers appear in July and August.

Drew's golden-eyed-grass, or Drew's
yellow-eyed-grass

Of the many yellow-flowered species of buttercup (*Ranunculus* spp.), one of the most common in our pine belt is **WATER-PLANTAIN BUTTERCUP (*Ranunculus alismifolius* var. *alismellus*)**. In the buttercup family (Ranunculaceae), this plant has erect stems that are about two inches to one foot high and simple, entire leaves. The flowers have five petals about one-fourth inch long and many stamens and pistils, which is typical of the buttercup family. It is often abundant in meadows

Water-plantain buttercup

and on wet banks from 4,500 to 12,000 feet in the San Jacinto and San Bernardino Mountains, Sierra Nevada, northern White and Inyo Mountains, North Coast Ranges, Klamath Mountains, and Cascade Range and then to Washington and Montana. It blooms in June and July.

Another Ranunculus is **ALPINE BUTTERCUP** *(Ranunculus eschscholtzii),* a low perennial that has stems to about 10 inches long and leaves with rounded lobes. The few flowers are terminal with bright yellow petals about half-an-inch

Alpine buttercup

long. This buttercup can be found about rocks and in meadows between 6,000 and 13,500 feet in the San Jacinto and San Bernardino Mountains, Sierra Nevada, and Great Basin areas of eastern California and through the Klamath Mountains and Cascade Range to Alaska and the Rocky Mountains. The flowering period is from July to August. (See "Whitish Flowers" for another Ranunculus species.)

California has many species of wild buckwheat *(Eriogonum)* in the buckwheat family (Polygonaceae), such as **TIBINAGUA,** or **NAKED-STEMMED ERIOGONUM,** *(Eriogonum nudum),* usu-

ally a few-branched perennial with basal leaves to about three inches long. The minute, clustered, yellow to white or pink flowers are less than one-fourth inch long and have a six-parted perianth. This plant is tremendously variable, having 13 different named varieties in California. Growing in dry, somewhat rocky places, it ranges from low elevations up to 12,500 feet through most of California north to Washington. Some varieties with flowers in solitary or paired (not clustered) involucres (cuplike structures below the flowers) occur in the mountains of southern California. The blooming period ranges from July to August.

Another wild buckwheat is **COMPOSITE ERIOGONUM (*Eriogonum compositum*),** a perennial with a rather woody branched base and basal leaves that are white woolly beneath. The stout flower stems are eight inches to over two feet tall and bear numerous pale yellow flowers about one-fourth inch long. This striking plant is found on dry, rocky walls and slopes below 8,000 feet in northwestern California and the Cascade Range and then to Washington and Idaho. It blooms from May to July.

Composite eriogonum

Sulfur flower

Another yellow-flowered wild buckwheat is **SULFUR FLOWER** *(Eriogonum umbellatum)*, many varieties of which are found in California. It is a perennial growing from a woody caudex and usually has several low stems that are leafy at the tips and bear umbels of simple or branched rays, which may or may not bear bracts near the middle. The flowers are in terminal, dense to open clusters and range from yellow to orange or reddish but are generally bright yellow with a reddish tinge. It grows in dry and rocky places below 12,000 feet in most California mountains and to the Rocky Mountains. The flowers appear between June and August. (See "Reddish Flowers" for other *Eriogonum* species.)

Golden eardrops

GOLDEN EARDROPS *(Dicentra chrysantha)* is in the poppy family (Papaveraceae)—but was formerly in the fumitory family (Fumariaceae)—and is closely related to bleeding heart (*D. formosa*) and steer's head (*D. uniflora*). Plants in this genus have flowers with four petals, the outer two petals are alike and have pouchlike structures at the bases, and the inner two petals are oblanceolate and joined at the tips. The only yellow-flowered member of the genus, this species has deeply dissected leaves between six and 12 inches long. The plant can be from two to over six feet tall, and with its yellow flowers, can be quite impressive. It is a fire follower, that is, it appears only after a fire or in disturbed sites and often disappears after a few years. The first two years after a fire, however, it can appear in great masses, creating a striking contrast to the blackened ground. It is found throughout most of California and blooms from April to September. (See "Reddish Flowers" for a pink *Dicentra* species.)

In the mustard family (Brassicaceae), with its acrid sap and four-parted flowers, is **AMERICAN WINTER-CRESS** *(Barbarea orthoceras)*, which has rather stout stems four inches to two feet tall. The basal leaves are often entire, but the stem leaves are divided. The pale yellow petals are to about one-fourth inch long, and the narrow seedpods are up to two inches. This winter-cress grows on stream banks, in springy places, and in meadows, largely between 2,500 and 11,000 feet in much of California, to Alaska and the Atlantic Coast, and in Asia. Its blooming period ranges from May to September.

Another crucifer (member of the mustard family) is **MOUNTAIN TANSY-MUSTARD (Descurainia incana),** a slender, pubescent annual with divided leaves. The stems are one to four feet high and usually branched above. The bright yellow petals are only one-eighth of an inch long, and the narrow seedpods are up to an inch. This tansy-mustard is found in dry, disturbed places between 5,000 and 11,000 feet in the San Bernardino Mountains, Sierra Nevada, Great Basin areas of eastern California, and Klamath Mountains northward to Washington and Alberta. It can be found blooming between May and August.

WESTERN WALLFLOWER (Erysimum capitatum subsp. perenne), another member of the mustard family, has four petals, and its ovary is up inside the flower. It is a short-lived perennial and four to 12 inches high, and its root crown is clothed with the remains of the old, spoon-shaped, obtuse leaves. The yellow petals are over half an inch long, and the ascending seedpods are two to three inches long. It is found in dry places from 6,500 to 13,000 feet in the Sierra Nevada, Cas-

Western wallflower

cade Range, and Klamath Mountains. The flowering period is from June to August.

CUSHION-CRESS (Draba lemmonii), also of the mustard family, forms a spreading, leafy cushion with leaves usually less than an inch long and leafless flower stems up to six inches high. The yellow petals are about one-fourth inch long, and the oval seedpods are one-fourth to slightly less than one-half inch in length. It occurs on gravelly and rocky slopes, in crevices, and on talus above 8,000 feet in the Sierra Nevada. It blooms in July and August.

The pitcher-plant family (Sarraceniaceae) is composed of insectivorous plants whose leaves are modified for catching insects and whose flowers nod at the ends of long stems. **CALIFORNIA PITCHER-PLANT,** or **COBRA-LILY,** **(Darlingtonia**

Cushion-cress

California pitcher-plant, or cobra-lily

californica) has each conspicuously veined leaf ending in a hooded tip with two hanging appendages. The sepals are yellow green, and the petals are purple and an inch or more long. This plant occurs in marshy and boggy places from 300 to 7,000 feet in the northern Sierra Nevada and Klamath Mountains into Oregon. The flowering season is from April to June.

The stonecrop family (Crassulaceae) is mostly succulent and has five-petaled flowers. An annual member of this family is **CONGDON'S SEDELLA** *(Parvisedum congdonii)*. It is only one to three-and-a-half inches high and diffusely branched. The tiny flowers have 10 stamens and five petals that are only one-twelfth inch long and spreading. It occurs in rocky places below 5,000 feet in the Sierra Nevada. A similar species, Sierra sedella (*P. pumilum*), has slightly longer petals that spread early but become erect in age. This species grows in rocky places and beds of vernal pools below 4,000 feet in the Sierra Nevada and North Coast Ranges, as well as in the Central Valley. Both species bloom from March to May.

A perennial member of the stonecrop family is **PACIFIC STONECROP (*Sedum spathulifolium*)**, a perennial with slender rootstocks and rather prominent rosettes of leaves. The flower stems are two to nine inches high and bear reduced leaves. The lanceolate yellow petals are long acuminate and

Pacific stonecrop

Lax live-forever, or spreading live-forever

up to one-third inch long. It occurs in rocky places up to 8,000 feet in the Transverse Ranges, Sierra Nevada, and Cascade Range, and in central and northwestern California and northward to British Columbia. The flowers appear between May and July. (See "Reddish Flowers" for another *Sedum* species.)

Also in the stonecrop family is **LAX LIVE-FOREVER,** or **SPREADING LIVE-FOREVER,** *(Dudleya cymosa),* a variable, more or less glaucous plant with mostly oblanceolate fleshy leaves one-half to seven inches long. The flower stems are to about a foot tall and bear reduced leaves. The bright yellow to reddish petals are about half-an-inch long. In the pine belt, this live-forever ranges to elevations of 9,000 feet and occurs in much of our mountainous areas from the San Bernardino Mountains northward. It blooms from April to June.

In the rose family (Rosaceae), two similar plants are ivesias (*Ivesia* spp.) and horkelias (*Horkelia* spp.), but ivesias have filiform filaments on the stamens, whereas horkelias have dilated filaments. A yellow-flowered species of the first genus is

GORDON'S IVESIA (Ivesia gordonii), which has a thick, woody caudex and leaves bearing 20 to 32 leaflets that are divided to the base. The small flowers have narrow petals, five stamens, and generally two to four pistils. This species is found in dry, rocky places between 6,000 and 11,500 feet in the Sierra Nevada, North Coast Ranges, Warner Mountains, Mount Eddy in Trinity County, and the Sweetwater Mountains in the eastern Sierra Nevada and then to Washington and the Rocky Mountains. The blooming period is July and August.

Cinquefoils (*Potentilla* spp.), also of the rose family, belong to a large genus. **SHRUBBY CINQUEFOIL (Potentilla fruticosa)** is the only woody species in the genus in California, and it is much branched and one to four feet tall. The leaflets are crowded on the divided leaves, and the inflorescence usually consists of one to five terminal, yellow flowers with round petals one-fourth to one-half inch long. Although the species is circumpolar, only the California plants have densely hairy fruit. It grows in moist places from 6,500 to 12,000 feet from Tulare and Inyo Counties northward and blooms from June to August.

DRUMMOND'S CINQUEFOIL (Potentilla drummondii) is a perennial with fairly erect stems up to two feet high that branch above and bear few leaves. The basal leaves have two

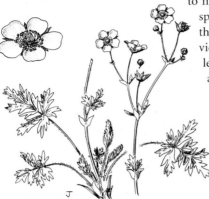

to five pairs of moderately spaced green leaflets, and the stem leaves are divided into one to five leaflets. Flowers are half-an-inch in diameter. This species is generally found in moist places from 3,500 to 12,000 feet in northwestern California, the Sierra Nevada, the Cascade Range, the Great Basin areas of eastern California, and then to British Columbia. A subspecies, brewer's cinquefoil *(P. drummondii* subsp. *breweri)*, has white-woolly leaves. The blooming period is July to August.

SLENDER CINQUEFOIL (Potentilla gracilis) is another perennial but has digitate, or palmate, leaves. They are green to hairy above and quite hairy and whiter beneath. The yellow petals are almost half-an-inch long. Quite variable, this plant is generally found in moist places below 11,500 feet and grows from the mountains of San Diego County to Alaska and South Dakota. It blooms from May to August.

Slender cinquefoil

Bigleaf avens

BIGLEAF AVENS (Geum macrophyllum) is a pinnately leaved perennial of the rose family and is characterized by the persistent hooked styles on the dry one-seeded pistils. The stems are bristly hairy and one to three feet high, and the compound leaves have very large, rounded, terminal leaflets. The petals are one-sixth inch or longer. This plant grows in moist places such as meadows from 3,500 to 10,500 feet in the San Bernardino Mountains, Sierra Nevada, Cascade Range, Great Basin areas of eastern California, in northwestern California, and to Alaska, eastern Asia, and Labrador. It blooms from May to August.

Lotuses (*Lotus* spp.), which have flowers typical of the pea family (Fabaceae), comprise a large group in California and usually have their flowers in umbels. **BROAD-LEAVED LOTUS *(Lotus crassifolius)*** is a perennial with stout stems one to five feet high and pinnately compound leaves divided into nine to 15 broad leaflets an inch or so long on either side of the central midrib. The inflorescence is an eight- to 20-flowered, one-sided umbel of yellow green flowers that may have purple blotches. The

Sierra Nevada lotus

corolla is about an inch long and forms a seedpod that can be up to three inches long. Growing on dry banks and flats from 1,000 to 7,000 feet, this lotus ranges from southern California to Washington and flowers from May to August.

Quite a different lotus is **SIERRA NEVADA LOTUS (Lotus nevadensis)**, a prostrate perennial with many slender wiry branches forming mats one to three feet across. The leaves have three to five appressed-hairy leaflets, and the umbels have three to 12 yellow to reddish flowers one-fourth to one-half inch long. It grows on dry sandy and gravelly slopes and benches from 3,500 to 8,500 feet in the mountains from San Diego County through most of montane California. It can be found blooming from May to August.

Narrow-leaved lotus

Another lotus is **NARROW-LEAVED LOTUS (Lotus oblongifolius)**, an erect or ascending perennial one to one-and-a-half feet high that has expanded stipules (appendages at the base of a leaf stem) and three to 11 leaflets to each leaf. The flowers are clustered at the ends of long peduncles (flower stems), and the corolla is about half-an-inch long. It is an inhabitant of wet places below 8,500 feet from southern California to Oregon. It flowers between May and September.

SCOULER'S ST. JOHN'S WORT (Hypericum formosum var. scouleri) is in the St. John's wort family (Hypericaceae). It has paired leaves dotted with translucent glands. With running rootstocks and erect bushy stems about one to two feet high,

Scouler's St. John's wort

this plant bears yellow flowers that are almost an inch in diameter. The numerous stamens are united into three groups. It is frequent in wet meadows and on banks from 4,000 to 8,000 feet in the mountains of southern California, Sierra Nevada, and Cascade Range and at lower elevations in the Coast Ranges from Monterey County northward to British Columbia and Montana. The blooming period is from June to August.

Among California's violets, a good many species are yellow or yellowish, such as **PINE VIO-LET** *(Viola lobata)*, a perennial in the violet family (Violaceae). It has an erect stem and can be over a foot tall. The leaves, mostly near the tips of the stems, are divided about halfway down into three to

Pine violet

seven lobes; occasionally they are entire. The deep yellow flowers have petals one-third to one-half inch long, and the two upper petals are purplish on the back. This species occurs on rather dry slopes in open woods from 500 to 7,500 feet in the Peninsular Ranges, Sierra Nevada, and Cascade Range, in northwestern California, and to southern Oregon and northern Baja California. The flowers appear from April to July.

MOUNTAIN VIOLET (Viola purpurea) is a variable perennial one to several inches high. The leaves are rounded to pointed, wedge shaped to heart shaped at the base (obcordate), and range greatly in size and pubescence. The petals are a deep lemon yellow, and the two upper petals are purplish brown on the back. This violet is found in dry places below 10,000 feet and ranges through most of the mountains of California to Oregon. It flowers between April and June.

SHELTON'S VIOLET (Viola sheltonii) has stems that project only a little above the ground and leaves divided into many linear lobes. The flowers are deep lemon yellow with dark veins, and the two upper petals are purplish brown on the back, as in pine violet (*V. lobata*) and mountain violet *(V. purpurea)*. It is a plant of open woods or brushy places between 2,500 and 8,000 feet and is found in much of montane California and northward to Washington. The blooming season is from April to July. (See "Bluish Flowers" for another *Viola* species.)

In the loasa family (Loasaceae) is **VENTANA STICKLEAF,** or **VENTANA BLAZING STAR,** *(Mentzelia congesta),* an annual with deeply lobed leaves cov‐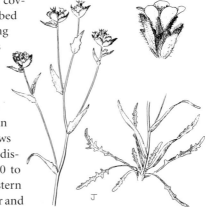ered with dense, barbed hairs that make them cling to clothing. The petals are pale yellow with an orange base and scarcely one-sixth inch long. There are numerous sta‐mens, and the fruit is an elongate capsule. It grows on dry burns and other dis‐turbed places from 5,000 to 9,000 feet in the eastern Sierra Nevada, Peninsular and western Transverse Ranges, and San Gabriel and Tehachapi Mountains. Ventana stickleaf flowers from May to July.

In the evening-primrose family (Onagraceae), one of the most conspicuous members is the true **EVENING-PRIMROSE** *(Oenothera elata* subsp. *hirsutissima).* The species itself is variable and is found in most parts of California from sea level to about 9,000 feet. This montane subspecies is about three to almost eight feet high, often with a reddish floral tube one to two inches long with four broad, notched, yellow petals one to two inches long and wide. It grows in moist places from 3,000 to 9,000 feet from the San Jacinto Mountains north‐ward. Flowers open toward sunset and wilt the next day when the sun becomes hot. The more coastal form is *O. elata* subsp.

Evening-primrose

hookeri, and it has the same common name as *O. elata* subsp. *hirsutissima.* Flowering is between June and September.

An evening-primrose of more local distribution is the peculiar **WOODY-FRUITED EVENING-PRIMROSE** *(Oenothera xylocarpa),* a stemless perennial that has densely soft-hairy, pinnately divided leaf blades one to three inches long and often spotted red. The flowers bloom in the evening hours and have

Woody-fruited evening-primrose

bright yellow petals about an inch long that age red. The woody capsule is four winged. This evening-primrose is a plant of dry benches under pines from 7,000 to 10,000 feet in the central and southern Sierra Nevada. It blooms from July to August.

PINNATIFID SUN CUP (*Camissonia tanacetifolia*) is a stemless perennial in the evening-primrose family that has very deeply lobed leaves and four-petaled yellow flowers that open in the day and can be almost an inch wide. The stigma is entire,

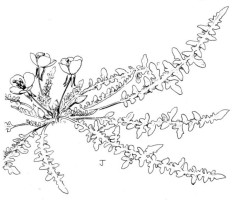

which distinguishes this genus from the above *Oenothera*, which has a four-lobed stigma. This sun cup is found in moist, sunny places from 2,000 to 8,000 feet in the Sierra Nevada, Cascade Range, and Great Basin areas of eastern California and to Washington and Idaho. A closely related species, northern sun cup (*C. subacaulis*), has leaves that can be entire or few toothed and ranges as far south as Tulare County. The flowering period for these two species ranges from May to August.

In the figwort family (Scrophulariaceae), one of the large groups is monkeyflower, or *Mimulus*, and one of the most common of the montane species is **MEADOW MONKEY-**

Meadow monkeyflower, or primrose monkeyflower

Larger mountain monkeyflower

FLOWER, or PRIMROSE MONKEYFLOWER, (Mimulus primuloides). The bright yellow flowers arise from rosettes or short tufts of three-veined leaves that range from glabrous to quite hairy. The corolla tube is less than an inch long, and each lobe of the lower lip usually has a reddish spot at its base. Often growing from stolons along the ground, the plants can form large mats in meadows and on wet, grassy banks from 2,000 to 11,000 feet. It ranges from the San Jacinto Mountains in southern California to Washington and flowers from June to August.

A particularly beautiful monkeyflower of the higher mountains is **LARGER MOUNTAIN MONKEYFLOWER** *(Mimulus tilingii)*, which has few-flowered stems and bright green leaves that can be over an inch long. The yellow, two-lipped flowers are one to two inches long with brown-spotted ridges in the throat. This plant occurs on wet banks from 4,500 to 11,000 feet in most of montane California except the South Coast Ranges, and then northward to British Columbia. It flowers from June to August.

Another smaller monkeyflower is **VARICOLORED MONKEY-FLOWER** *(Mimulus whitneyi)*, which is remarkable because it has flowers of various colors: pale yellow with maroon blotches and lines or purple with similar paler areas. The plant is only one-half to five inches high, the leaves are less than an inch long, and the almost sessile flowers are about half-an-inch long. This species of monkeyflower grows in gravelly places from 5,000 to 11,000 feet in the southern Sierra Nevada. The flowering period is from June to August. (See "Reddish Flowers" for a pink *Mimulus* species.)

Another member of the figwort family is **YELLOW OWL'S-CLOVER** *(Castilleja tenuis)*, a slender-stemmed annual that is from a few inches to over a foot high, is more or less hairy, and has narrow leaves, the uppermost of which are three to five lobed. The flowers are about half-an-inch inch long and borne in a terminal spike. It is found in meadows from 3,000 to 9,000 feet, rarely in San Bernardino and San

Yellow owl's-clover

Diego Counties, and more frequently from Glenn and Tulare Counties northward to Alaska and Idaho. It flowers from May to August. Other similar species differ in technical details. (See "Reddish Flowers" for another *Castilleja* species.)

SIERRA PODISTERA (*Podistera nevadensis*) is in the carrot family (Apiaceae). It is a compact, stemless perennial with a crown of fibrous sheaths and tufted short leaves with three to seven divisions. The flower stems are to about an inch long and bear heads of yellow flowers. It is generally found above timberline from 10,000 to 13,000 feet in the San Bernardino Mountains, northern and central Sierra Nevada, and White and Inyo Mountains. Flowering is from July to September.

The sunflower family (Asteraceae), with its heads of small tubular disk flowers, narrow petal-like ray flowers, or both, has many yellow-flowered species. **MOUNTAIN WYETHIA,** or

Sierra podistera

MOUNTAIN MULE-EARS, *(Wyethia mollis),* is in many ways quite like the sunflower itself. It is a resin-dotted perennial that is densely woolly when young. Its stems are one to almost two feet tall, and its basal leaves are eight to 16 inches long, whereas the stem leaves are shorter. The stems bear one to four flower heads about an inch across that have only five to 11 ray flowers but multiple disk flowers. This mule-ears occurs on dry, wooded slopes and in rocky openings from 4,000 to 11,000 feet from the northern and central Sierra Nevada through the Klamath Mountains and Cascade Range to southeastern Oregon and Nevada. Related species are less woolly or may have basal and stem leaves that are more similar to each other. The flowering period ranges from May to August.

Mountain wyethia, or mountain mule-ears

Arrow-leaved balsamroot

ARROW-LEAVED BALSAMROOT (*Balsamorhiza sagittata*), also a member of the sunflower family, is a heavy-rooted perennial with stems about one to two feet tall. The basal leaves are large and widely triangular, but the upper leaves are linear to oblanceolate. The one to few flower heads are about two to four inches across. This plant grows in deep, sandy, open places between 4,300 and 8,300 feet in the Sierra Nevada and Great Basin areas of eastern California and north to Canada and the Rocky Mountains. Other species may have narrower leaves. Flowering is from May to July.

Related to the preceding plants is **CALIFORNIA HELIANTHELLA (*Helianthella californica*),** a perennial with leafy stems up to two feet tall. The narrow leaves are five to 10 inches long and not more than one-and-a-half inches wide. The heads are solitary and long stemmed with nine to 21 yellow petal-like ray flowers and numerous small, central disk flowers of the

California helianthella

same color. Growing on grassy slopes below 8,500 feet, it has several varieties and ranges from the Sierra Nevada through the Cascade Range, Klamath Mountains, and North Coast Ranges to Oregon. It also occurs in the southern Peninsular Ranges. It blooms from May to September.

CALIFORNIA CONE-FLOWER *(Rudbeckia californica)* also belongs to the sunflower family. It has a leafy, unbranched stem two to several feet tall with a single showy head on each peduncle. The leaf blades are four to 12 inches long. The center of the flower head is made up of many greenish yellow disk flowers, and it elongates into a rounded column an inch or more high, giving this genus its common name of "cone-flower." The ray flowers are yellow. There are three varieties, and all are occasional in moist places up to 8,500 feet in the Sierra Nevada, Klamath Mountains, and North Coast Ranges to Oregon. They

California cone-flower

Common madia

flower in July and August. More common is black-eyed Susan (*R. hirta* var. *pulcherrima*), which was introduced from the Midwest and can be distinguished by its dark purple disk flowers.

COMMON MADIA (*Madia elegans*), another member of the sunflower family, is an erect annual that is branched and glandular above and hairy below and has numerous, narrow leaves. The heads have five to 21 yellow ray flowers that often have a maroon blotch at the base. The disk flowers are yellow or maroon. This madia is common on rather dry slopes and at the edges of meadows up to 11,000 feet throughout much of California. There are four subspecies, three of them having conspicuous ray flowers from one-fourth to one-half inch long, but in *M. elegans* subsp. *wheeleri*, they are only about one-sixth inch long. The flowering period for most of the subspecies is June to August, but *M. elegans* subsp. *vernalis* flowers in the spring.

Bolander's madia, or Bolander's tarweed

The *Madia* genus is one of many genera that make up a group of plants called tarweeds because of the ill-smelling viscid glands on the plants. **BOLANDER'S MADIA,** or **BOLANDER'S TARWEED,** *(Madia bolanderi)* is a perennial with woody rootstocks and a stem about two to four feet high that is more or less bristly below and with many stalked glands above. The op-

posite leaves are linear, hairy, and four to 12 inches long. The flower heads are few and in an open inflorescence, and each have eight to 12 outer ray flowers and 30 to 65 inner disk flowers. This madia grows in damp mountain meadows or along streams from 3,500 to 8,500 feet in the Sierra Nevada,

Cascade Range, Klamath Mountains, and North Coast Ranges and then to Oregon. It blooms from July to September.

WHITNEYA (Whitneya dealbata) was named for John D. Whitney, to whom Mount Whitney, California's highest peak, is dedicated. The plant is perennial with one to a few erect stems up to 14 inches tall. The whitish, densely woolly, three-nerved leaves are two to four inches long, but smaller above. The solitary or few heads are long stemmed and have yellow ray flowers up to an inch long and numerous yellow disk flowers. Found in light, moist soils of open hillsides and forests, this plant is uncommon, occurring from 4,000 to 8,000 feet in the northern and central Sierra Nevada and Cascade Range. It blooms from June to August.

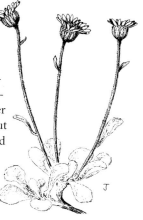

Hulseas (*Hulsea* spp.) are members of the sunflower family that are found mainly in the mountains. **PUMICE HULSEA (Hulsea vestita)** is a perennial with dense rosettes of white-woolly, entire or toothed leaves and several stems, each ending in a single flower head. The ray flowers are linear, about one-third inch long, and yellow tinged with red or purple. Sometimes reduced to a dwarf form at higher altitudes, as in Parry's hulsea (*H. vestita* subsp. *pygmaea*), this hulsea grows in

Red-rayed hulsea

sandy or gravelly soils from 4,000 to 13,000 feet in the Sierra Nevada, Transverse and Peninsular Ranges and Desert Mountains. There are six subspecies, some of them quite local in their distribution. Most of them bloom between June and August.

RED-RAYED HULSEA (*Hulsea heterochroma*) is a robust, sticky-hairy, and heavy-scented perennial with several erect stems bearing several oblong leaves that are toothed and up to eight inches long. The 30 to 60 petal-like ray flowers are yellow to reddish purple. It is found from 1,000 to 8,000 feet in forest openings from the San Jacinto Mountains northward to Santa Clara County in the Coast Ranges, as well as in the Sierra Nevada. It blooms from June to August.

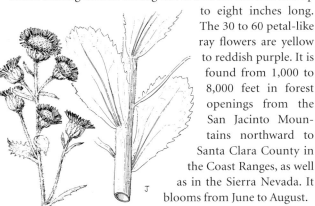

ALPINE GOLD *(Hulsea algida)* is a glandular hairy, disagreeably odorous perennial with few to many one-headed, leafy stems four to 16 inches high. The leaves are basal, more or less coarsely toothed, and up to four inches long. The heads of 25 to 60 ray flowers and numerous disk flowers are two or more inches across. This plant is found on rocky peaks from 10,000 to 13,000 feet and ranges from Mount Whitney to Mount Rose, Nevada. Dwarf

Alpine gold

hulsea *(H. nana)* is found in the Lassen Peak and Mount Shasta areas and has pinnately lobed leaves. It occurs on volcanic talus and ranges from the Cascade Range to the Modoc Plateau and Washington. The flowering period for both alpine gold and dwarf hulsea ranges from July to August.

BIGELOW'S SNEEZEWEED, or **BIGELOW'S SNAKEWEED,** *(Helenium bigelovii)* is also in the sunflower family and has a stem from one to about four feet high that can be branched above or throughout. The lower leaves have short, winged stems, and the upper leaves are sessile and decurrent, that is, they continue to run down the stem below their point of

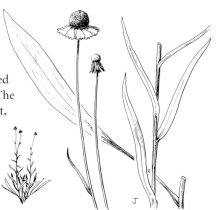

attachment, forming a wing along the stem. The flower heads are long stemmed with 14 to 20 showy ray flowers up to almost an inch long surrounding the central, rounded mound of disk flowers. Although these disk flowers have yellow tubes, their lobes can be yellow, red, brown, or purple, thus coloring the center of the flower head. It is common in moist meadow-like places to about 10,000 feet in most of California and to Oregon. Flowering is from June to August.

GOLDEN-YARROW, or **WOOLLY-SUNFLOWER,** *(Eriophyllum lanatum)*, also of the sunflower family, is one of the most variable species in California and has eight named varieties. It is

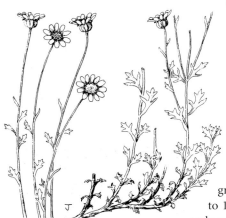

usually perennial, erect or decumbent from a woody base, and one to about three feet high, and it has a persistent or partially deciduous woolly covering. The leaves are usually toothed or divided. The flower heads are solitary or in open groups, each with eight to 13 yellow ray flowers and numerous disk flowers. It is rather common in brushy places below 13,000 feet from southern California to Washington and Montana. The flowering period ranges from April to August for the different varieties.

DISSECTED BAHIA *(Bahia dissecta)*, another plant in the sunflower family, is a biennial or short-lived perennial up to four feet high with leaves divided into linear or oblong segments.

The flower head consists of 10 to 20 ray flowers and numerous disk flowers. The species grows on gravelly open slopes and dry, rocky ridges from 6,000 to 8,600 feet in the San Bernardino and Santa Rosa Mountains and to Wyoming and Texas. It blooms in August and September.

Pyrrocomas (*Pyrrocoma* spp.) are another large group in the sunflower family and are related to the asters (*Aster* spp.). **ALPINE PYRROCOMA,** or **ALPINE GOLDEN-ASTER,** *(Pyrrocoma apargioides)* has several stems two to seven inches long and basal tufted leaves one to four inches long. The heads are usually solitary and have 11 to 40 ray flowers one-eighth to one-

Alpine pyrrocoma, or alpine golden-aster

third inch long. It is found in open, rocky places and in meadows from 7,500 to 12,000 feet in the Sierra Nevada and the Inyo and White Mountains. This plant blooms from July to September.

WHITESTEM GOLDEN-BUSH *(Ericameria discoidea)* is closely related to alpine pyrrocoma *(Pyrrocoma apargioides)* above. A low shrub four to 16 inches high, it has densely white-woolly branches and numerous leaves about an inch long. The heads have no ray flowers but are made up of 10 to 26 yellow tubular disk flowers. It grows on rocky, mostly open slopes, often on talus above timberline, between 9,000 and 12,000 feet in the Sierra Nevada and the Warner and Sweetwater Mountains and to Oregon and the Rocky Mountains. It flowers from July to September.

Another compact shrub in this genus is **BLOOMER'S MACRONEMA** *(Ericameria bloomeri),* which has a woody trunk up to an inch thick. The numerous leaves vary from almost filiform, or threadlike, to oblanceolate. There are few to many flower heads in a cluster, each having one to five ray

Bloomer's macronema

flowers, and four to 13 disk flowers. It is found in sandy or rocky places between 3,500 and 9,500 feet in the Sierra Nevada, Cascade Range, and Modoc Plateau and to Washington. It blooms from July to October.

A third species of *Ericameria* is **CUNEATE-LEAVED ERICAMERIA (*Ericameria cuneata*),** a deep green, spreading, much-branched shrub with crowded, balsamic-resinous leaves, which may be wedge shaped to roundish (particularly in southern California) and up to an inch long. The flower heads are in compact clusters with one to three ray flowers (or sometimes none) and seven to 70 smaller tubular disk flowers. It can be found on slopes and cliffs, often in crevices in granitic rocks, up to 9,500 feet in the Sierra Nevada and the mountains of southern California. It blooms late in the year, from September to November.

Cuneate-leaved ericameria

Goldenrod (*Solidago* spp.) is a sunflower family member represented in the higher mountains by **NORTHERN GOLDENROD** *(Solidago multiradiata),* a perennial with erect stems two to 20 inches high. The basal leaves are up to five inches long, and the upper leaves are smaller. The few to many heads each have 12 to 18 ray flowers and 12 to 35 disk flowers. It inhabits sunny, rocky or grassy places mostly from 8,500 to 12,500 feet and occurs in the Sierra Nevada, Inyo and White Mountains, Cascade Range, and Klamath Mountains and then to Alaska and Siberia, as well as east to Labrador. It flowers between June and September.

Northern goldenrod

Rabbitbrushes (*Chrysothamnus* spp.), also in the sunflower family, are very common throughout our western states, especially in subalkaline places. **YELLOW RABBITBRUSH (Chrysothamnus viscidiflorus)** has numerous small flower heads consisting of three to 13 disk flowers and no ray flowers. The heads are surrounded below by five rows of leaflike phyllaries. The twigs are not woolly as in some species, and the rounded shrub is usually less than three feet high with linear leaves. There are four named varieties occupying dry, open places between 3,000 and 13,000 feet from southern California to British Columbia and Montana. This plant blooms from July to September.

PARRY'S RABBITBRUSH (Chrysothamnus parryi) is a shrub less than three feet high with twigs covered in feltlike wool. The yellow heads of disk flowers are on long branches. This species too has several varieties but in general grows on mountain sides and flats from 2,000 to 12,000 feet from southern California to Siskiyou and Modoc Counties. Like other species of rabbitbrush, Parry's rabbitbrush blooms in late summer and fall, from July to September.

COLUMBIA CUTLEAF, or **COLUMBIA HYMENOPAPPUS,** *(Hymenopappus filifolius var. lugens)* is a southern member of the sunflower family. It is a perennial from eight inches to two feet

Columbia cutleaf, or Columbia hymenopappus

tall with a basal rosette of leaves dissected into filiform divisions. The three to eight flower heads per stem are made up of many yellow disk flowers about one-fifth inch long. There are no ray flowers. The plant is found on dry slopes from 4,000 to 9,000 feet in the Transverse and Peninsular Ranges of southern California and to Utah and New Mexico. It blooms from June to August.

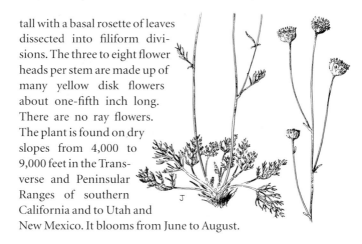

Arnicas (*Arnica* spp.) are somewhat aromatic or glandular perennial herbs in the sunflower family that grow from rhizomes or caudices and have mostly paired leaves. The rather large, flat-topped heads are generally solitary, but occasionally there are several. **MEADOW ARNICA (*Arnica chamissonis***

Meadow arnica

subsp. *foliosa)* is one to almost three feet high and
somewhat hairy with five to
10 pairs of leaves two to 12
inches long. The petal-like
ray flowers are one-half to
three-fourths of an inch
long, and there are numerous small disk flowers in the
center. It is common in
meadows and moist places
from 6,000 to 11,500 feet in
the San Bernardino Moun-
tains, Sierra Nevada, Warner,
Inyo, White, and Klamath
Mountains, and Cascade
Range, thence to Alaska and
the Rocky Mountains. It
flowers from July to August.

GREEN-LEAVED RAILLARDELLA (Raillardella scaposa), another plant in the sunflower family, is a low perennial with
mostly basal green leaves that
can be over six inches long.
The leafless, or scapose,
stems are up to 16 inches
tall and glandular, and each
bears a flower head with nine
to 44 disk flowers and zero to
seven ray flowers. It grows in
dry, rocky places and at the
edges of meadows from
6,500 to 11,500 feet in the
Sierra Nevada and in Ore-
gon and Nevada, flower-
ing in July and August.

Silky raillardella, or silver mat

Dwarf mountain butterweed

SILKY RAILLARDELLA, or **SILVER MAT, (Raillardella argentea)** is a silky-woolly perennial up to six inches high, and the head bears an involucre about half-an-inch long. It has seven to 26 disk flowers and no ray flowers. Growing in dry, rocky places from 7,000 to 13,000 feet, it can be found in the San Bernardino Mountains and Sierra Nevada and northward to Oregon. Its gray leaves distinguish it from the Sierran green-leaved raillardella *(R. scaposa)* above, which has green, glandular leaves and sometimes short ray flowers, and from the related, but less common, scabrid raillardella *(Raillardiopsis scabrida),* which has branched stems and can be found in the North Coast Ranges and Shasta County. All three species bloom between July and August.

One of the largest groups in the sunflower family are groundsels, ragworts, or butterweeds *(Senecio* spp.), and they can often be recognized by their involucres, which are made up largely of equal bracts with much shorter ones at the base. **DWARF MOUNTAIN BUTTERWEED (Senecio fremontii)** is a perennial from a branching base, and it has slender, decumbent stems and rather fleshy, rounded leaves. The heads are solitary at the ends of the branches, and the involucres are almost half-an-inch high. The eight yellow ray flowers are one-fourth to one-third inch long, and there are up to 40 disk flowers. It grows on rocks from 8,500 to 12,000 feet in the San Bernardino Mountains, Sierra Nevada, and Cascade Range and to British Columbia. It is also found in the Great Basin areas of eastern California. It flowers in July and August.

A common montane butterweed is **ARROWHEAD BUTTERWEED (Senecio triangularis),** which has elongate-triangular leaves and a flat-topped inflorescence of flower heads that consists of eight ray flowers and up to 40 disk flowers. It can be found in wet meadows and on stream banks from 3,000 to 11,150 feet in the San Jacinto, San Bernardino, and San Gabriel

Arrowhead butterweed

Mountains, Sierra Nevada, Klamath Mountains, Cascade Range, and Modoc Plateau and northward to Alaska and the Rocky Mountains. The flowers appear from July to September.

Quite a different butterweed is the misnamed **RAYLESS ALPINE BUTTERWEED *(Senecio pauciflorus)*,** which can actually have up to 13 ray flowers, although it often has none. It has up to 40 disk flowers. The stem is four to 20 inches high, and the mostly basal leaves are more or less toothed. The several flower heads are umbel-like and somewhat reddish to orange. This butterweed grows in wet meadows from 8,000 to 11,500 feet in the Sierra Nevada and to Alaska and Labrador. It blooms in July and August.

Rayless alpine butterweed

Alpine rock butterweed

A final montane butterweed is **ALPINE ROCK BUTTERWEED** *(Senecio werneriifolius)*, a perennial with tufted basal leaves that are white woolly to greenish, smooth, and entire or nearly so. The several stems are four to eight inches high, almost leafless, and bear one to six flower heads, each with eight to 13 ray flowers (or occasionally none) that are one-fourth to one-third inch long. There can be up to 40 disk flowers. This species is found in dry, rocky places from 10,000 to 13,000 feet in the northern and central Sierra Nevada and White and Inyo Mountains. It flowers from July to August. Similar species occur in wetter areas and are more widespread geographically.

Members of the false-agoseris genus *Nothocalais* are dandelion-like plants with flower heads composed of only ray flowers. **ALPINE LAKE–AGOSERIS** *(Nothocalais alpestris)* is a stemless perennial with a basal rosette of mostly toothed or deeply lobed leaves and solitary flower heads at the tips of the peduncles (flower stems). The outer and inner phyllaries are the same length, but the outer are wider and usually purple dotted. It is found only infrequently and occurs in meadows and on open slopes from 6,000 to 8,000 feet in El Dorado County in the northern Sierra Nevada, and in the Klamath Mountains, then northward to Washington. It blooms in July and August.

SHAGGY HAWKWEED *(Hieracium horridum)* is also composed of only ray flowers. It is a perennial and has one to several shaggy-hairy stems four to 14 inches tall, with spatulate to oblong leaves. The inflorescence is open and has many small heads, each bearing six to 15 bright yellow ray flowers. It is common in dry, more or less rocky places from 4,500 to 11,000

feet in the San Jacinto Mountains, Sierra Nevada, Great Basin areas of eastern California, Klamath Mountains, and Cascade Range and to Oregon. It blooms in July and August. (See "Whitish Flowers" for another *Hieracium* species.)

WESTERN HAWKSBEARD (*Crepis occidentalis*) is a final yellow-flowered member of the sunflower family here, and it, also, has only ray flowers. It is up to 16 inches tall and has a dense, gray, woolly covering. The leaves are four to 12 inches long, toothed to lobed, and gradually reduced up the stems, which are several branched from the base or middle, each main branch having 10 to 30 clustered flower heads. The yellow ray flowers are almost an inch long. The species is highly variable and occurs in dry, rocky places from 2,000 to 9,000 feet in the San Bernardino and Tehachapi Mountains, western Transverse Ranges, Sierra Nevada, Cascade Range, Klamath Mountains, North Coast Ranges, Mount Hamilton in the San Francisco Bay Area, and Great Basin areas of eastern California and then to Washington and Wyoming. Flowering is from June to August.

GLOSSARY

Anther The pollen-forming portion of the stamen (male reproductive part of a plant).

Appressed (leaf, etc.) A leaf that is pressed against the stem. Other structures may also be appressed, such as leaf hairs against a leaf.

Basal (leaf) A leaf found at or near the base of a plant.

Blade The expanded, generally flat portion of a leaf, petal, or other structure.

Boreal (species) A term referring to plants native to high latitudes of the northern hemisphere or lower latitudes at high elevations; short for *circumboreal*.

Bract A small, leaflike or scalelike structure usually subtending a flower or cone.

Calyx The outer whorl of flower parts consisting of sepals; the collective term for sepals.

Carpel A subunit of the pistil; it may be one or many; fused or free.

Caudex The short, sometimes woody, vertical stem of a perennial plant. The plural is *caudices*.

Cauline (structure) A structure that is borne on a stem and therefore not basal.

Cleft (petal or leaf) A petal or leaf whose margin is indented.

Compound (leaf) A leaf that is composed of two or more leaf-like parts.

Corm A short, thick, underground bulblike stem without scales as in crocuses or gladiolas.

Corolla Whorl of flower parts immediately inside or above the calyx.

Crenate (leaf) A leaf that has a scalloped margin in between shallow, rounded teeth.

Decompound (leaf) A leaf that is divided more than once.

Decumbent (plant) A plant that mostly lies flat on the ground but with stems or flowers that curve upward.

Disk flower The rayless cluster of generally five-lobed flowers (as in the sunflower family).

Filament The anther-bearing stalk of a stamen.

Filiform (structure) A thread-shaped structure, as found in stems, leaves, or flower parts.

Floret A single flower and its immediately subtending bracts (as in the grass family).

Follicle A dry, generally many-seeded fruit from a single pistil, opening on one side and when ripe along a single suture.

Frond A leaf (often referring to ferns).

Funnelform (corolla) A corolla that widens from the base more or less gradually through the throat.

Glabrous (leaf) A smooth, hairless leaf.

Glaucous A plant structure covered with a whitish or bluish, waxy or powdery film.

Glume One of the usually paired bracts at the base of a grass spikelet.

Herbaceous (plant) A plant that is not woody.

Hispid (leaf) A leaf that is covered with straight, stiff hairs.

Inferior ovary An ovary growing below the calyx.

Inflorescence A group of flowers and associated structures.

Involucre A group of bracts more or less held together and subtending a flower or a group of flowers.

Keel The two lowermost, fused petals of many members of the pea family.

Lemma The lower, generally larger of two sheathing bracts subtending a flower (as in the grass family).

Lobe A bulge, as on the margin of a leaf or petal.

Margin The leaf or petal edge.

Midrib The main vein running lengthwise down the middle of a leaf.

Node The position on an axis or stem from which structures such as a leaf arise.

Ovary The ovule-bearing, generally wider lower part of a pistil that normally develops into a fruit.

Ovule A structure within the ovary containing an egg.

Palea The upper and generally smaller of two sheathing bracts subtending a flower (in the grass family).

Palmately compound (leaflets, veins) Leaflets or veins of a leaf that radiate from a common point.

Palmatifid (veins, leaflets) Leaflets or veins of a leaf that are very deeply divided into lobes spreading like the fingers from the palm of a hand.

Parasite A plant that benefits from a physical connection to another living species, often harming the host.

Pedicel A flower stalk; support of a single flower.

Peduncle The stalk of a flower, fruit, or inflorescence.

Pendent (leaves, petal) Leaves or petals that droop or hang.

Perianth The structure of a flower comprising the calyx and corolla.

Petal An individual member of the corolla.

Phyllary The bract of an involucre that subtends a flower head (in the sunflower family).

Pinna The primary division of a pinnate leaf or fern frond, or leaflet. The plural is *pinnae*.

Pinnate (leaf or fern) A leaf or fern that has similar parts arranged on opposite sides of an axis.

Pinnately compound (leaf) A compound leaf made up of pinnate leaflets.

Pinnatifid (leaf or fern) A pinnately divided leaf or fern in which the divisions are very deep.

Pinnule A secondary pinna.

Pistil The female organ of a flower, comprising the ovary, style, and stigma.

Pollen The fertilizing, dustlike powder produced by the anther.

Pubescence A covering of soft hair or down.

Ray flower The outer, often three-lobed, showy petal-like flower at the edge of a flower head (in the sunflower family).

Reflexed (leaf, petal) A leaf or petal that is abruptly bent or curved downward or backward.

Reniform A structure that is kidney-shaped.

Rhizome An underground, more or less horizontal stem, as in cordgrass.

Rosette A radiating cluster of leaves generally at or near ground level.

Runner A stolon or any basal branch that is disposed to root.

Salverform (corolla) A corolla that has a slender tube and an abruptly spreading throat.

Saprophyte A plant that lives on dead organic matter.

Scape A leafless floral axis or peduncle arising from the ground.

Sepal An individual member of a calyx.

Sessile (flower, leaf) A flower or leaf that does not have a stalk.

Seta Bristle or bristle-shaped body. The plural is *setae.*

Sorus A cluster of sporangia (reproductive cells) in ferns. The plural is *sori.*

Spikelet The smallest aggregation of florets and subtending glumes (in grasses).

Sporangium The spore-producing organ in nonseed plants. The plural is *sporangia*.

Spore The minute, dispersing, reproductive unit of nonseed plants (ferns and fern allies).

Stamen The male reproductive structure of a flower with a stalk-like filament and pollen-producing anther.

Stigma The part of a pistil on which pollen is deposited, generally terminal and elevated above the ovary.

Stipe The leafstalk of a fern or of a pistil.

Stolon A runner or basal branch that is disposed to root.

Style The stalk that connects an ovary to the stigma.

Superior ovary An ovary growing above the calyx.

Ternately compound (leaf) A leaf that is compounded into three parts, such as a clover leaf.

Truncate A leaf or other structure that is severely attenuated, appearing "lopped off" at the base or tip.

Umbel An inflorescence (flower) in which three to many pedicels radiate from a common point, as in parsley.

Viscid (plant) A plant that has a sticky surface.

Whorl A group of three or more structures (e.g., leaves, flower parts) at one node.

ART CREDITS

Photographs credited to the California Academy of Sciences Collection are also credited to their individual photographers.

Line Illustrations

DICK BEASLEY 6

TOM CRAIG 163 (bottom),195

RODNEY CROSS 93, 181, 188 (bottom)

PETER GAEDE 28, 29, 31, 40 (top, bottom), 44, 53 (top), 56, 62 (top), 67 (top, bottom), 69, 70, 71 (bottom), 72, 78, 79, 84, 94, 95 (top), 107, 111, 115 (bottom), 130 (top, bottom), 142, 150, 151 (top), 162 (top), 177 (bottom), 180, 183, 185, 198 (top)

JEANNE R. JANISH 24 (top, bottom), 25, 26 (top, bottom), 27, 34 (top, bottom), 39, 41, 45, 46 (top, bottom), 47 (top, bottom), 48, 49, 51, 53 (bottom), 59, 60 (top, bottom), 62 (bottom), 63 (top, bottom), 64 (top, bottom), 66, 71 (top), 75 (top, bottom), 76 (top, bottom), 77, 82 (top, bottom), 83 (top, bottom), 87, 88, 89, 90, 91, 92 (top, bottom), 95 (bottom), 96, 97, 98 (top, bottom), 99, 100 (top, bottom), 101 (top, bottom), 103, 104 (top, bottom), 105 (top, bottom), 108, 109, 112, 113 (top, bottom), 114 (top, bottom), 115 (top), 118, 120, 121 (top, bottom), 123, 124 (top, bottom), 125, 126, 129 (top, bottom), 134, 135, 136, 137 (top, bottom), 139, 145, 148, 149 (top, bottom), 151 (bottom), 153, 154, 157, 158, 159, 162 (bottom), 163 (top), 164, 165, 167, 169, 172 (top, bottom), 174, 175, 176, 177 (top), 179, 182, 187 (top, bottom), 188 (top), 189, 190, 192, 193 (top, bottom), 194, 196, 197, 198 (bottom), 199, 201, 202 (top, bottom), 203, 204, 205 (top, bottom), 206, 207, 208, 209, 210 (top, bottom), 211, 212, 213 (top, bottom), 214, 215, 218, 220, 221 (top, bottom)

PAULA NELSON AND BILL NELSON 20, 21

Color Photographs

SHERRY BALLARD 25

ALBERT P. BEKKER 80–81

BROTHER ALFRED BROUSSEAU 73, 75, 92, 109, 127 (top), 135, 138 (bottom), 147 (top), 165 (bottom), 179, 191, 202, 212 (bottom)

CALIFORNIA ACADEMY OF SCIENCES 25, 27 (bottom), 30 (bottom), 39, 43 (top), 45 (top, bottom), 50 (top, bottom), 52, 55 (bottom), 58 (top), 65 (top right), 70 (top), 72, 78, 80–81, 88, 89, 90, 98, 102 (top, bottom), 103, 106 (top, bottom), 108, 111, 112, 116, 117 (top, bottom), 121, 126, 129, 130, 132 (top, bottom), 133, 134, 136, 138 (top), 143, 147 (bottom), 148, 151, 156 (top), 157, 158, 159, 160, 161 (bottom), 164, 166, 168 (top, bottom), 170–171, 176, 177 (bottom), 178, 180, 181, 185, 186, 189 (top, bottom), 190, 192 (top), 195 (top, bottom), 206, 209, 211, 212 (top), 214, 218, 219 (top)

GERALD AND BUFF CORSI 45 (top), 58 (top), 70 (top), 166, 176

DR. G. DALLAS AND MARGARET HANNA 168 (bottom)

LORRAINE ELROD 50 (top)

WILLIAM T. FOLLETTE 54, 66, 68 (top), 74, 146, 161 (top), 216 (bottom)

JOHN GAME ii–iii, vi, xii–1, 22–23, 26, 27 (top), 28, 29, 30 (top), 32–33, 35, 36, 37 (top, bottom), 38, 41, 42, 43 (bottom), 48, 51, 55 (top), 56, 57 (left, right), 58 (bottom), 59, 60, 61, 64, 65 (top left, bottom), 68 (bottom), 69, 70 (bottom), 76, 77, 85 (top, bottom), 86 (top, bottom), 91, 93, 94, 95, 96, 100, 101, 110, 119 (top, bottom), 120, 122, 125, 127 (bottom), 128, 131, 140–141, 142, 144, 145, 149, 150, 152 (top, bottom), 153, 154, 155, 156 (bottom), 165 (top), 167, 173, 174 (top, bottom), 175, 177 (top), 183, 184 (top, bottom), 192 (bottom), 197 (top, bottom), 199, 200 (top, bottom), 201, 203 (left, right), 204, 207, 215, 216 (top), 219 (bottom)

BEATRICE F. HOWITT 126, 195 (top), 206

WALTER KNIGHT 185

J. E. (JED) AND BONNIE MCCLELLAN 133, 134

DONALD MYRICK 108, 130, 209

ROBERT POTTS 45 (bottom), 121, 138 (top), 143, 159, 186, 195 (bottom), 211

GLADYS LUCILLE SMITH 72, 102 (bottom), 117 (top), 161 (bottom)

JULES STRAUSS 89

CHARLES THOREAU TOWNSEND 43 (top)

CHARLES WEBBER 27 (bottom), 39, 50 (bottom), 52, 55 (bottom), 65 (top right), 78, 88, 90, 98, 102 (top), 103, 106 (top, bottom), 111, 112, 116, 117 (bottom), 129, 132 (top, bottom), 136, 147 (bottom), 148, 151, 156 (top), 157, 158, 160, 164, 168 (top), 170–171, 177 (bottom), 178, 180, 181, 189 (top, bottom), 190, 192 (top), 212 (top), 214, 218, 219 (top)

INDEX

Page references in **boldface** refer to the main discussion of the species

ABOUT THE AUTHOR
AND EDITORS

Philip A. Munz (1892–1974) of the Rancho Santa Ana Botanical Garden was professor of botany at Pomona College, serving as dean for three years. Phyllis M. Faber is general editor of the California Natural History Guides. Dianne Lake is rare plant committee cochair and unusual plants coordinator for the California Native Plant Society, East Bay Chapter.